YANKEE CITY SERIES

VOLUME
IV

PUBLISHED ON THE

Richard Teller Crane, Jr., Memorial Fund

The Yankee City Series Is Dedicated to
CORNELIUS CRANE

Each volume in the Yankee City Series is complete in itself.

THE
SOCIAL SYSTEM
OF THE
MODERN FACTORY

THE STRIKE:
A SOCIAL ANALYSIS

BY
W. LLOYD WARNER
AND
J. O. LOW

NEW HAVEN
YALE UNIVERSITY PRESS
LONDON · GEOFFREY CUMBERLEGE · OXFORD UNIVERSITY PRESS
1947

Copyright, 1947, by Yale University Press

Printed in the United States of America

All rights reserved. This book may not be reproduced, in whole or in part, in any form (except by reviewers for the public press), without written permission from the publishers.

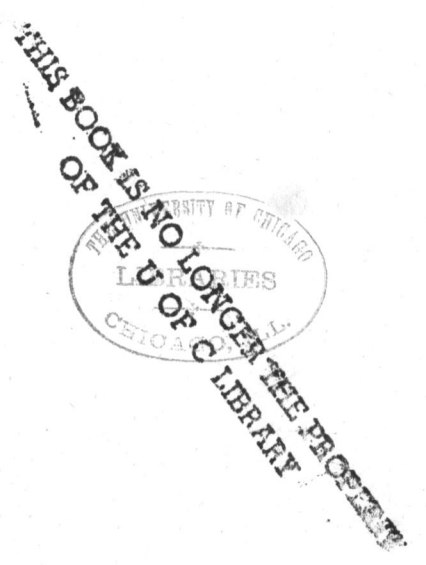

This volume is dedicated to

John Dollard

YANKEE CITY RESEARCH STAFF

Writer W				Analyst A	Field Worker F		
W. Lloyd Warner	W	A	F		Allison Davis	A	F
Leo Srole	W	A	F		Elizabeth Davis	A	F
Paul S. Lunt	W	A	F		Burleigh Gardner	A	F
J. O. Low	W	A	F		Alice Williams	A	
Eliot Chapple		A	F		Mildred Warner	A	
Buford Junker		A	F		Joseph Weckler	A	
Solon Kimball		A	F		Dorothea Mayo		F
Marion Lee		A	F		O. S. Lovekin		F
Conrad Arensberg		A	F		Gwenneth Harrington		F
Robert G. Snider		A	F		Hess Haughton		F

ACKNOWLEDGMENTS

THE research staff of the Yankee City Series are indebted to many people for aid in the inception, course of research, and publication of this work. We cannot thank all of them here.

Above all we wish to express our appreciation to the citizens of Yankee City who generously gave their knowledge of the community and maintained a coöperative interest in our work. We are grateful to the Committee of Industrial Physiology of Harvard University for sponsoring and financing the field research. Professor Elton Mayo and Dean Wallace Brett Donham of the Graduate School of Business Administration of Harvard University have contributed incalculably to the study by their wise guidance, insight, and understanding.

The entire series is dedicated to our generous benefactor, friend, and colleague in social anthropology, Cornelius Crane, as an inadequate recognition of his deep and sympathetic interest in our work and of his consuming concern for the problem, the nature of man.

The authors are particularly indebted to Professors Edwin B. Wilson, Carl R. Doering, and Earnest A. Hooton, all of Harvard University, and Professor Samuel A. Stouffer of the University of Chicago, for advice and assistance on statistical problems.

We wish to thank Professors John Dollard and George Peter Murdock, of Yale University, for their interest in the publication of these volumes and for their unerring aid in the editing of the manuscripts.

Miss Alice Marsden White and Mrs. Mildred Hall Warner have contributed their skill in the preparation of the manuscripts for publication. We wish to thank our friend and colleague, Dr. Mark A. May, and his staff of the Institute of Human Relations at Yale University, for the encouragement they gave us by their critical and sympathetic interest.

To John Dollard of the Institute of Human Relations, to whom this volume is dedicated, we owe a very special debt of gratitude for his recognition of the significance of the scientific problems we attacked and for his help in the solution of many of them. The searching questions he asked us and the generous acclaim he gave our research have been deep sources of scientific and spiritual strength to all of us.

The authors are deeply indebted to Marion Lee for the use of her excellent field document on the strike and case histories about some of the personalities in the shoe industry. We have used her materials extensively to provide us with some of the evidence for part of our conclusions.

CONTENTS

LIST OF TABLES	xiii
LIST OF CHARTS	xiv
PREFACE	xv
I. THE STRIKE—WHY DID IT HAPPEN?	1
1. The Strike	1
2. Why Did It Happen?	4
II. PRELUDE TO CONFLICT	8
1. The Depression as a Social Setting and Factor for the Strike	8
2. Grievances	24
III. THE NATURAL HISTORY OF A STRIKE	31
1. Battle Strategy	31
2. The Unions Fight Management for Control of the Workers in the Community	33
3. The Union Takes Control and Attacks Management	42
4. Peace Negotiations—The Third Period	50
IV. FROM CLIPPERS TO TEXTILES TO SHOES	54
1. The Industrial History of Yankee City	54
2. From the Cobbler's Bench to Assembly Line—A Shoe History	59
3. The Strike and the Evolving Social and Economic Systems	63
V. THE BREAK IN THE SKILL HIERARCHY	66
1. Skill and Status	66
2. Forms of Social Control in the Factory	67
3. The Break in the Skill Hierarchy	73
4. Social Effects of the Break	80
5. The Strike and the Break in the Skill Hierarchy	87

Contents

VI.	Wages and Worker Solidarity	90
	1. Women, Wages, and Solidarity	90
	2. Ethnic Groups and Solidarity	92
	3. Wage Factors	98
VII.	Yankee City Loses Control of Its Shoe Factories	108
	1. Big City Men Take Over	108
	2. Horizontal Extensions of the Factories	121
VIII.	Managers and Owners, Then and Now	134
	1. The Managers of Men Were Gods	134
	2. Little Men and Aliens Run Things Now	140
	3. Structural Analysis of the Old and New Managers' Status	150
IX.	The Workers Lose Status in the Community	159
	1. A Coming Industrial Proletariat?	159
	2. The Causes of the Yankee City Strike	172
X.	Blue Print of Tomorrow—General Conclusions	181
	1. Economic Change and Social Class in America	181
	2. Species Behavior and a Planned Society	189
	3. World Implications of the Extension of the Economic Hierarchy	193

Appendices

1. The Shoe Factory as a Social Structure — 197
2. The Workers and Their Departments — 200
3. Frequency of Membership of the Workers — 217
4. The Range of Family Membership — 222
5. Ethnic Workers' Memberships — 225
6. Social Characteristics of the Shoe Workers — 227

Index — 237

TABLES

1. Classification of Jobs, and Groups of Jobs, Showing the Number of Male Workers Employed in Them and the Difference between Their Average Hourly Earnings and the Average for All Male Operatives (in 1934) — 94
2. Jobs for Which Operatives Are Paid More than the Factory Average for Men — 101
3. Jobs for Which Operatives Are Paid Less than the Factory Average for Men — 101
4. Sample Tabulation of Extreme Range of Social Participation — 165
5. Associational Membership—Extreme Range — 167
6. Clique Membership—Extreme Range — 169
7. Association Memberships Per Capita — 217
8. Clique Memberships Per Capita — 218
9. Professions of Religious Faith by Percentage — 219
10. Political Participation by Social Classes — 221
11. Total Number of Family Members by Social Stratification — 223
12. Percentage of Memberships in Families Which Varied in Range from the One-Class Norm, Showing Variations by Ethnic Groups between Shoe Operatives and Total Yankee City Population — 224
13. Memberships Per Individual, Shoe Operatives Compared with Total Adult of Three Lower Classes, Shown by Social Class, Ethnicity, and Totals — 226
14. The Social Characteristics of Shoe Workers — 228
15. The Social Characteristics of the Technological Workers in Different Departments — 232

CHARTS

I. The History of the Differentiation of the Yankee City Shoe Industry *opposite* 64

II. Permitted Scope of Activities in Jobs of Different Relative Statuses 68

III. Distinction between Types of Skill in Jobs and the Degree or Extent to Which Proficiency May Be Developed by Operators within the Varied Classifications 75

IV. The Result of the Leveling of Technological Jobs in the Shoe Factory 81

V. Vertical Extension of the Managerial Hierarchy 111

VI. Shoe Manufacturing in Yankee City 115

VII. Horizontal Organization in Shoe Factories 122

VIII. The Change from Craft to Industrial Unionism and Its Relation to Division of Labor and Destruction of the Skill Hierarchy in Technological Jobs 125

IX. The Positional System of Yankee City 163

X. Social Mobility in the School; Blocked Mobility in the Factory 184

XI. The Formal Structure of a Large Shoe Manufacturing Enterprise 198

XII. Social Stratification of the Shoe Operatives as Compared with Total Yankee City Community 230

XIII. Percentage Figures of Social Stratification of Shoe Operatives by Sex 231

PREFACE

THE "Yankee City Series," of which the present work is the fourth volume, will be complete in six volumes. Each deals with a significant aspect of the life of a modern community as it has been recorded and analyzed by the combined and coöperative labors of a group of social anthropologists. The same techniques and viewpoints applied by them to the study of societies of simpler peoples are here subjected to empirical testing in a concrete case study in modern American society. The town chosen (for reasons given in Volume I) was an old New England community.

The first volume, *The Social Life of a Modern Community*, by W. Lloyd Warner and Paul S. Lunt, describes in detail the cultural life of the community, emphasizing particularly the way in which these people have been divided into superior and inferior classes. It also presents the reader with an interpretation of the techniques, methods, and conceptual framework used in the research, a summary of the findings, and a general orientation.

The second volume, *The Status System of a Modern Community*, by W. Lloyd Warner and Paul S. Lunt, gives a detailed description and careful analysis of the social institutions of this community. It shows how our New England subjects lived a well-ordered existence according to a status system maintained by these several social institutions.

The third volume, *The Social Systems of American Ethnic Groups*, by W. Lloyd Warner and Leo Srole, is a detailed study of the social life of a number of ethnic groups, including the Irish, French Canadians, Jews, Armenians, and Poles; it explains how they maintain their old cultural traditions but at the same time undergo social changes which make them more and more like the larger American community.

The present volume, *The Social System of the Modern Factory*, by W. Lloyd Warner and J. O. Low, is specifically concerned with the study of the social organization of the modern factory. It shows not only how industrial workers co-operate in producing manufactured goods, but also how they fit into the larger community.

The fifth volume, *American Symbol Systems*, by W. Lloyd Warner, deals with the conceptual processes which Americans use when they think about themselves and their own behavior. It analyzes the myth and ritual as well as the secular behavior of the members of Yankee City.

The concluding volume, *Data Book for the Yankee City Series*, by W. Lloyd Warner, supplies additional data for those who wish to examine the more detailed aspects of the subjects treated in the other volumes.

I

THE STRIKE—WHY DID IT HAPPEN?

1. *The Strike*

THE best of all possible moments to achieve insight into the life of a human being is during a fundamental crisis when he is faced with grave decisions which can mean ruin and despair or success and happiness for him. In such crises men reveal what they are and often betray their innermost secrets in a way they never do and never can when life moves placidly and easily. If this is true for the study of men as individuals, it applies even more forcefully to the study of men in groups. It is when hell breaks loose and all men do their worst and best that the powerful forces which organize and control human society are revealed. We learn then, if ever, why groups of men must do the things they do and be the things they are. It is in these moments of crisis that the humdrum daily living of the thousands of little men going to work with their lunch boxes and the prosaic existence of the big man in the top office reveal themselves as human dramas of the utmost significance; more importantly, behavior in such crises tells us the meanings and significance of human society.

We have selected such a crisis in the relations of management and workers in a factory of Yankee City for analysis. The study of this dramatic conflict illuminates the normal position of the factory in the community and the relations of management with labor.

On a cold March day in the worst year of the depression all the workers in all the factories of the principal industry of Yankee City walked out. They struck with little or no warning; struck with such impact that all the factories closed and no worker remained at his bench. Management had said their workers would never strike because the workers of Yankee City were sensible and dependable, and had proved by a long peaceful history that they would always stay on the job. Union men outside the city said the Yankee City workers would not strike because Yankee City had never been and could not be organized and, furthermore, the shoe workers of Yankee City were obstinate and "always stupid enough to play management's game." Many of the workers had told us that there would be no strike. Most of the townspeople, from the Breckenridges and Blaisdails of Hill Street to the O'Malleys and Debenskis in the clam flats, said Yankee City workers would never strike. But foreigners and Yankees of ten generations, men and women, very old and very young, Jews and Gentiles, Catholics and Protestants—the whole heterogeneous mass of workers left their benches and in a few hours wiped out most of the basic production from which Yankee City earned its living. Not only did they strike and soundly defeat management, but they organized themselves, joined an industrial union, and became some of its strongest members.

The strike occurred almost in the very year of the three-hundredth anniversary of the founding of Yankee City and the beginning of the shoe industry. Shoemaking had always been important in the economy of the town, but it was not until near the end of the nineteenth century that it achieved a place of supreme importance. From the beginning, shipping, shipbuilding, fishing, and the other trades of the sea had dominated Yankee City's economic existence and set their mark on the community. When the New England shipping industries disappeared, Yankee City turned from the sea and sent its salesmen to the prairie and mountain states to sell its manufactured goods to make the profits necessary for the establishment and continuance of its factory system. It was then that the textile manufactures moved into the lead, but throughout the whole period shoemaking contributed significantly to the economic life of the city. Yankee City's shoe workers and owners throughout this time were known everywhere in the country for the excellence of their products.

Although the economy of the city went through revolutionary changes, the social superstructure which guided and maintained the lives of the citizens remained very much what it had been at the end of the War of 1812. The physical city stretches in a thin rectangle two miles inland from the harbor along the bank of a large river. Here, when the field study was made, lived 17,000 people. They were distributed from the river bottoms and clam flats back to the high ground of Hill Street. The people of high status, some of them the descendants of those who made their fortunes in the sea trade, lived on this broad, elm-lined avenue. The people of lowest status, many of whom could trace their ancestry through long lines of fishermen to the city's founding, lived in Riverbrook on the clam flats. Between the two were the Side-streeters who, appropriately enough, occupied a middle-class status.[1]

The upper class of Hill Street is composed of two levels: the "Old Families," who can trace their aristocratic position through an ancestry of many generations, and the "New Families," who have but recently achieved high status. In the latter group are several families who "got their money out of shoes." The upper-middle class, the "pillars of society," and the lower-middle classes, the top of the "common man" group, are very much like such people wherever they are found in the United States or, for that matter, in all English-speaking countries. They are the conservatives, who, dominated by a "Protestant ethic," maintain and often control the moral order of the city. Below them is the upper-lower class composed of the "poor but honest workmen" in the factories. At the bottom are the "broken-down Yankees," often called the "Riverbrookers," who also work in the factories and do a moderate amount of clamming and fishing.

Scattered throughout the status system from the lower-upper ("New Family") class to the lower-lower ("Riverbrookers") are the descendants of the Irish and, at somewhat lower levels, the French Canadians, Jews, Poles, Greeks, and other ethnic groups who began settling in Yankee City in the 1840's and continued until 1924.[2] Each group has its own social system which preserves an increasingly small stock of the ancestral culture while relating its population to the larger world of Yankee City. The Yankees are the most powerful group in the city, but the ethnics each year increase their power and prestige while they shed their variant mores and accept those of the dominant Yankees.

1. See W. L. Warner and P. S. Lunt, *The Social Life of a Modern Community*, "Yankee City Series," Volume I; and *The Status System of a Modern Community*, "Yankee City Series," Volume II.
2. See W. L. Warner and Leo Srole, *The Social System of American Ethnic Groups*, "Yankee City Series," Volume III.

2. Why Did It Happen?

ALL groups in Yankee City were involved in the strike; the income of most of them was directly or indirectly derived from the shoe factories. Men everywhere in the city asked themselves, when the strike occurred, why such a thing could have happened to the people of Yankee City. Each man had his own answer, which usually revealed more about the life and status of the speaker than about the causes of the strike. But, despite the ethnocentric and egocentric sentiments which dominated the people's reasoning, in each answer we always found at least a molecule of truth. Sometimes it was hard to determine just what was implied.

The people on Hill Street declared that the causes of the strike could be found only by a search far back into the history of Yankee City. Characteristically, they took the position of the historian and believed contemporary life could only be explained as the last domino to fall. They said, "How can one understand the strange behavior of those odd people in the shoe business today? How different the owners were in the old days!" One old lady declared, "Our todays are made out of yesterday. In fact, sometimes I think we are yesterday."

One of the local managers said it was the depression. "No one," he said, "could sell shoes in a buyer's market unless the workers would take further cuts in their wages." Many of the workers said it was the low wages—"A man couldn't feed his family on what he was paid." Other workers said it was a plot. Johnny, the Greek shoe cutter, declared it was "because the rich were not satisfied with what they had but were trying to take everything away from the workers and turn all of the workers into slaves." Sometimes when such things were said the representatives of management insisted the workers' words gave clear evidence of a "Red plot" and that "if it had not been for union agitators there'd have been no trouble and no strike." Each man, owner and worker and townsman, spoke his own brand of economic determinism. Their words would have been dear to the ears of those scientists who interpret all human phenomena in terms of the economic factor.

Each of these speakers, however, forgot that there had been serious depressions before and that there had been no strike in

Yankee City. Each of them forgot that there had been low wages before and that there had been no unions. Each forgot, too, that there had been strikes when wages were high and times were said to be good. Although these economic arguments supplied important and necessary reasons for the strike and the unionization of the workers, they were insufficient to explain the strike and why unionization occurred.

Other citizens of Yankee City were less clear than these men about the reasons for the strike. Some of them said, "It's the goddam foreigners." Of these who spoke, many were Irish Catholic, particularly of the middle class, and when they spoke they were thinking about Greeks, Poles, Armenians, and other more recently arrived peoples. Other men said, "It's the goddam foreigners." And they thought of these same ethnic groups but added the Catholic Irish and the French Canadians. Sam Jones, Tim Green, and other Riverbrookers said it was the goddam foreigners, and they meant the owners and managers of the factories from New York, or, for that matter, anywhere outside of Yankee City. Other workers narrowed the circle and did not bother to go beyond the Jews. They said, "It's those lousy New York Kikes. Jews always ruin everything they get into—once a Kike gets into anything, decent people have to get out."

Everyone in management and labor agreed that the strike could not have happened "if only fine men like Caleb Choate, Godfrey Weatherby, and William Pierce were running the factories now to take care of things," for they would know what to do and they would know how to act. But Caleb, Godfrey, and William were dead. They had long since taken up their residence in the Elm Hill Cemetery.

These are but a few of the common answers given to the question of why the strike had to happen in Yankee City.

During the strike our interviewers were spread throughout the town to observe what the people said and did. We knew many of these people intimately and well, for we had been living in the city and studying it for several years when the strike occurred.[3] We sat in with management; we talked with the

---

3. The field work supplying the evidence for this volume was done in two periods, the first in 1930–35, and the second, ten years later, in the fall of 1945. The materials for the entire research included in the six volumes covered the years of 1930–35; most of the facts bearing directly on the factory

workers, the union organizers, the tradesmen, and the government arbitrators. All of them told us why the strike had happened. Each told but part of the truth; no one knew all of it.

The secret of why the Yankee City workers struck and of why men in other cities strike seems to us to lie beyond the words and deeds of the strike. The answers can be found only in the whole life of the community of which the workers and owners are but a part. The answers of the economic determinists or of the historians, while important, are not sufficient.

If social science is to be of any worth to us it must be capable first of all of adding significance and meaning to human behavior which will give us deeper insight into human life and explain more fully than common-sense knowledge why human beings act the way they do. Science necessarily solves problems. To solve them it must know what questions are involved. Let us reëxamine the questions implied in the statements of the Yankee City townsmen in a more explicit and pointed manner to determine whether we can learn what happened in this industrial crisis and to see if such knowledge about the strike can tell us about other similar crises in American life.

The immediate questions are basic to the whole problem; but answering them leads us into the more important fundamental questions about the nature of our industrial society. We will endeavor in this book to try to give at least partial answers to some of these larger questions.

The first questions we must answer about the strike are:

1. In a community where there had been very few strikes and *no* successful ones, why did the workers in *all* of the factories of the largest industry of the community strike, win all their demands and, after a severe struggle, soundly defeat management?

2. In a community where unions had previously tried and failed to gain more than a weak foothold and where there had never been a strong union, why was a union successful in separating the workers from management?

and the shoe workers were accumulated in 1934 and 1935. The strike occurred in 1933.

All necessary facts for the basic generalizations in this book were re-checked by the field work of 1945. We found no essential differences between the two periods; the events intervening greatly strengthened the evidence for our original conclusions.

3. Why was the union successful in maintaining the organization despite the intense and prolonged efforts of management to prevent unionization and to halt the continuation of the shoe union?

4. Why did Yankee City change from a non-union to a union town?

Before going further, we must examine the events which precipitated the strike. This evidence will give us the basic arguments of the economic determinists.

II

PRELUDE TO CONFLICT

1. The Depression as a Social Setting and Factor for the Strike

FOR approximately a year after the 1929 crash, the increasing unemployment in Yankee City did not attract widespread public notice. Local charitable organizations carried their increased burdens as best they could. But in November 1930, the local newspaper urged the city to do something for its unemployed. The coöperation of private citizens and welfare agencies was urged in an editorial which ran as follows:

A meeting should be called of all the employing interests in the city, merchants and professional men, who are interested to a man in the welfare of the city and its people, the leaders of Chamber of Commerce work, the City Welfare Department, and the leaders in other welfare agencies of the city. At this meeting a survey should be made of conditions; then, having determined the conditions, seek the means to remedy them.

The Yankee City *Herald* waged an editorial campaign for several weeks, urging the unemployed to register with the Chamber of Commerce. It also asked for subscriptions for the unemployed and carried free advertisements for those who needed jobs, stating applicants' abilities, number of dependents, etc.

The various charitable and fraternal organizations in Yankee City appointed a joint committee to investigate possible solutions of the local unemployment problem. The suggestions of this committee were that efforts should be made on a strictly private basis by individual employers and citizens:

There was a general discussion of the situation, many favorable suggestions being made by those present. It was the frequently expressed opinion that employment was what was needed and that any funds raised be utilized for furnishing employment rather than outright charity, it being the belief of the discussionists that the worthy unemployed prefer work rather than charity.

There were a number of individual responses to this tacit appeal; not in jobs, however, for there were few jobs. But a barber offered to cut the hair of children of the unemployed, merchants donated goods to be distributed to the needy, and so on. At this time, too, the factories of the community, including the shoe factories, adopted a policy of giving part-time jobs to several individuals instead of one full-time job to one individual. This practice of spreading work later became a sore point with shoe operatives who claimed that the factories either were taking advantage of workers by paying out less in total wages than they had for similar work before staggering the jobs, or were giving workers no consideration, making all workers come every day and wait around the factory for several hours to get in an hour or two of work.

At this time, the general attitude among the upper classes was that the depression was only a temporary setback; prosperity was "just around the corner." Workers who had lost their jobs or feared they would be laid off next did not, however, share this optimism. Sam Dixon[1] (upper-lower class), an unemployed worker who was a spokesman for the lower-class point of view at this time, said: "The times are terrible. First big money came in (from the outside), then some little hinky-dinky shops closed up. We all said, 'What the hell, they don't matter!' Pretty soon as more of them closed up they began to matter—now everything is closed up." Sam was an expert shoe worker and was expressing his and other workers' distrust of owners who were strangers—his greater sense of security when local industry was locally owned and managed. He continued: "There's a little hinky-dinky place down the road that makes fire-doors. The owner died, there are no heirs, so that's through. The silver plant is working four days a week; silver has gone to hell. Weatherby and Pierce has shut down, but they have a little

1. See "Yankee City Series," I, 193–196, for a profile of Dixon and his upper-lower-class family. The reader will find it helpful to familiarize himself with Chapter VII, "Profiles from Yankee City," in Volume I before reading the next two chapters. The profiles provide more personal details about the individuals mentioned here.

corner in Bronstein's factory. Bronstein's is doing good business. It makes shoes for the ABC chain stores. Jones and Jackson is being run by some Yids and they're manufacturing shoes for a chain store. The comb shop is on quarter time. The only people who are making any money now are the clam diggers. They all used to be looked down on as just Riverbrookers in the old days, but they make good money—from thirty-five to forty dollars a week."

Here was a world gone topsy-turvy, where the Riverbrookers were making good money digging clams "when decent self-respecting guys couldn't even get jobs." Sam went on to say that even the Riverbrookers' rock of security was crumbling before the deleterious effect of chlorinated water on clams and the "bootlegging" of Maine clams. As Sam saw the situation, "there's no security anywhere, any more, for anybody, and I'm telling you, and you mark my words, it ain't a Nigger but a Kike that's in the woodpile."

At this moment, when the lower classes were getting badly frightened, they were further shocked by the closing of one of the city's manufacturing plants. "The story is simple," Sam told an interviewer. "This company made wooden heels for ladies' shoes. During the fall the company had been hiring about eighty men a day and forty at night. Those bastards on Hill Street, where the shop was, began yelling about all the noise. They got the law out and made the company stop working nights. A hearing was held, and an injunction obtained restraining the company from operating at night. They got sore and shut down the whole damn thing and got ready to move their equipment and the plant to another town. A hundred and twenty men were thrown out of work and when you count their families that meant a total of two hundred and fifty to four hundred individuals."

Protests of the citizens over these events were immediately forthcoming. Numerous letters appeared in the paper; the city, as a corporate entity, was berated for allowing the factory to be closed; the slogan, "Old Yankee City offers new opportunities," was howled down as "asinine and stupid." Merchants, banks, and churches were among the objectors. But the factory stayed closed. This incident brought out the antagonisms of lower-class workers to the upper-uppers and lower-uppers who "caused" the shutdown. John Donahoe, shoe worker, said: "The Mark Company has been driven out of town by a bunch of old crabs. They said they were disturbing their sleep. When they do get a good company here with a good payroll, who want to be here, why they go and drive them out."

Feelings ran particularly high against the prime mover among the objectors, a man who taught school in Boston. Because of his profession he was suspected by the workers of being a "Bolshevik." As one who had a good job outside of Yankee City, he was criticized for taking jobs away from local residents. Tim O'Malley[2] said: "Why don't the Fahertys and their kind go somewhere else if our town is too noisy for them? It's got to the point Yankee City ain't good enough for them any more."

Tim's wife, Annie, leaned over the back fence and told Mrs. Dugan, whose husband had lost his job "because them snobs don't want to wake up from forty years of sleeping," that it was all the fault of "people like the Fahertys, they want to turn Yankee City into an estate like the English did to Ireland. Ah, yes, before long there'll be a bailiff around to throw us out of our houses, just like Ireland. Then the Fahertys and the Starrs will be happy."

The local paper at this time represented upper-class opinion and tried to gloss over the closing of the wooden-heel mill by reporting that the mill officials said they had not abandoned the local plant but would reopen again as soon as they had a market for their product. The paper did not explain why a factory which had had to work day and night suddenly had no business. Obviously this story was an attempt on the part of the paper to quiet the workers' growing antagonism to the upper classes.

In December 1930, Thomas Brown (upper-middle), pillar of society and molder of public opinion for the three higher classes, said: "The outlook for local business is bright. I'm sure the depth of the depression has been reached. Improvement can be expected soon." He went on to state, among other optimistic portents, that the local shoe manufacturers "with a few exceptions have recently received many orders which will keep business up to normal during January and February. Other in-

2. See "Yankee City Series," III, 9–27, for profiles about the "Shanty Irish" O'Malleys (upper-lower) and the "Lace-curtain" Fahertys (lower-upper).

dustries I know are operating part-time; but the owners are optimistic for the future." The tone of opinion (upper-upper to upper-middle) at this stage of the depression continued to express the increasingly desperate hopefulness of these classes, who were really beginning to doubt the temporary nature of the setback but were as yet unable to face the prospects of a long depression. The day after Christmas, the *Herald* headlined: "Prosperity looms as unemployment commences to fade." On that day it is probable that a fifth of the breadwinners of the town were out of work.

In the early months of 1931, the upper classes began to hedge a little from the position of uncompromising cheerfulness. The appeal then made was to individuals: they were to make jobs; they were to buy and make prosperity. "Industrial conditions," said Mr. Brown, "are improving throughout the country. It's inevitable that this condition will soon be in New England, but in the meantime there are many Yankee City workers seeking employment for the care of their families. We must call on the generosity of the people of Yankee City to care for this situation." He spoke scathingly of those who had "lost the will to purchase," because of depression propaganda, and exhorted everyone to "buy now." The assumption was that if everyone would do his bit all would soon be well and the wheels of industry would again turn.

Similar exhortations were being made at this time under the leadership of President Hoover's Emergency Committee for Employment in every community of the country. The aim in each case was for the local community to solve its own unemployment problems. Mr. Brown prevailed on his friend, the owner of the solidly respectable paper, to point out "to people of our kind" what their duties as citizens of Yankee City were:

Do not think of unemployment in terms of millions of people out of work in this broad land of America.

Think of unemployment as a few people out of work within a stone's throw of you—your own neighbors. Maybe one or two out of every twenty in your vicinity. That, after all, is the unemployment problem. Put these people back to work and that ends the unemployment problem in your city. Don't worry about other communities.

. . .

Now, what about you? Now, for instance, is the time to make additions, improvements, repairs, to have odd jobs done around your home, to increase the worth of your property while materials are low priced. . . . to give a neighbor a job . . . and you are doing as patriotic a thing as any man can do. You are doing a constructive thing, a profitable thing, and a friendly thing . . . The only useful money is money at work—put some of yours to work.

The local government of Yankee City tried hard: emergency employment was found for 295 men in the tree and highway departments and in repairing and painting the schools. The latter work temporarily took care of the painters and carpenters, nearly all of whom had been on the relief rolls since the onset of the depression. An unsuccessful attempt was made to raise extra taxes for the street repair in order to make more jobs for the jobless.

But the battle, along the two lines of private and local governmental job-making, was a losing one. The curtailment of industrial activity continued; the unemployed became more numerous; and both private citizens and the local community exhausted odd-job possibilities as well as funds with which to pay for them. Families whose wage earners were unemployed lived on savings, if they had any; workers who were still employed took several cuts in pay.

During 1931 the efforts of the community to handle the unemployment situation turned into a frantic drive to raise funds. The charitable organizations assumed increasingly heavy loads as private savings were depleted. There was a determined effort to use their funds in the most efficient manner by unifying all the community efforts. A meeting of all the welfare and charitable organizations was called, and a majority voted to use the Community Welfare as a clearing house. Alexander French[3] (upper-middle), active in local philanthropy, said afterwards: "The Community Welfare is tied much closer to

3. See Volume I, pp. 152–155, for his profile.

City Hall now than it ever was before, and the purpose of the meeting was to get the people together. There've been a lot of complaints, especially from the ministers, and the idea was to make them collaborate and feel a sense of civic pride."

Thus an effort was made to coördinate the relief activities of all interested private individuals, charitable organizations, and the local government. Many of the churches and small private organizations turned over their funds to the Community Welfare Organization. But by June 1932 the load had become too heavy to carry. Private charity had practically dried up.

At this time all cases that appeared to have no early chance of recovering their economic independence were turned over to the Public Welfare (the charitable organization maintained by the city treasury). The load carried by this agency before the depression had averaged less than eighty families, but by the end of 1932 it had risen to four hundred families.

The burden was too much; local public relief collapsed as had local private relief before it. It was at this point that federal relief appeared in several forms to take up the burden that had exhausted all local resources.

Before national relief appeared on the scene, however, the fears and antagonisms raised among the workers by their precarious position had helped precipitate the shoe strike in the early spring of 1933.

There can be no question about the salutary effects of such federal agencies as the CCC and the CWA on the morale of Yankee City citizens of the lower class. As we shall show in tracing the emotional tensions which developed during the depression and culminated in the shoe strike, by the end of 1932 the working classes in Yankee City had become bitter and were in fear of actual starvation. All the resources of the local community had failed to provide work or even to supply adequate food for families whose wage earners were without jobs through no fault of their own. There was talk to the effect that the government was merely a tool of the rich and something had better be done quickly to make radical changes in it.

But federal funds did what local finances were inadequate to do: they provided jobs that staved off starvation and even preserved members of the lower classes from the excessive loss of social status which accompanied acceptance of charity, however meager, from local agencies. The rôle played by the federal government in the latter part of the depression was viewed very favorably by a Yankee City citizen more qualified than most to compare its efforts with those of local organizations. This man was, and had been for some time, head of the Public Welfare in Yankee City. He also became the local administrator for CWA and ERA when those federal agencies were formed. His statement shows clearly the precarious position of the lower classes in Yankee City at the time the federal government stepped into the relief picture, and shows how essential was federal intervention in local relief since, for the first time in its entire history, Yankee City was helpless to cope with the problems arising from an economic crisis. He said: "You'd be surprised if you knew the people who came in here. Why, last week one of our local ministers came in and asked me for help. I didn't like to see him go to the Welfare, so now he's working for ERA with a pick and shovel. I gave him ten dollars out of my own pocket and put him to work. Now that fellow can keep his self-respect.

"I don't take any more new cases into the Welfare than I can help. I put them on the ERA. A man doesn't lose his self-respect when he has a job. I put my desk on this side of the room so I can see out of the window over there and see the people waiting to come in and see me. You know there's a lot of difference in the look on their faces. They look happier and more confident since the ERA and CWA."

This last paragraph is particularly revealing. The administrator, himself, was obviously happier now that he could give jobs—he had moved his desk so that he could see the applicants in the waiting room where before he had not wanted to embarrass them or himself any more than necessary. He went on: "Last Tuesday I saw 360 people here and yesterday I saw 240. . . . We have two doctors, a minister, lawyers, business men, people you'd never believe could be on our books. When they come in we have to start from the bottom up because they wait until everything is used up before they come to us. Too proud, you know. We start them right at the bottom, give them shoes, clothing, underclothing, shirts and food. Since the government has started making mattresses I don't know how many calls I've had for them. Why, some of these people ain't got nothing. You remember, I told you about some of those houses where the people broke up the furniture to burn it up and keep warm. People don't realize how bad things are. I know I didn't until I took this job and went around and saw for myself."

Here the administrator was revealing the utter breakdown of local efforts to care for the indigent. He continued in the same vein: "St. Anthony's Guild folded up a year ago. The Father said they couldn't support the people they'd helped for years, old people. So he came to me and told me about it. Both the Protestant and Catholic churches have fallen down on charitable contributions. They ain't doing nothing. The Community Welfare, one of our biggest charities, took care of over two hundred cases two years ago. Now they take care of about sixty-five."

The administrator saw federal efforts to care for the indigent in sharp contrast to the chaotic results of local efforts, and put the contrast into words: "It makes me tired to hear some of these people talk and kick about the way Roosevelt is handling it. When I make these talks before different organizations and clubs some of them get up afterwards and blame the government for the increase in tax rates. They seem to think it's because of the Welfare. I lose my temper. Can't help myself. There may be better programs, but nobody seems to have thought of it yet and until they do this one seems to be doing the job. I'm not saying that there may not be something better, but I'm tired of hearing people crab about this one. You know, all these [local] charities have fallen down and all the people that they used to take care of have come to us, so it's all dropped into our lap."

Local charities and individuals, in other words, had finally given up completely; unemployment had become an accepted thing as had the community's inability to care for the unemployed. Even while some complained at the intervention of the federal government, everyone who had studied the problem with any care at all knew that Yankee City could not preserve its traditional independence in this situation and that it had to depend on the larger society for the support of many of its citizens.

But it was before federal aid came to Yankee City, when

possible starvation confronted the lower classes, that the shoe operatives in the community struck. Let us go back to 1931 and 1932, when the working classes were beginning to express open antagonism toward "the rich"—the upper classes and, particularly, the absentee owners of local factories. It will be recalled that at this time the upper classes themselves, and the local newspaper which represented essentially their point of view, were still determinedly optimistic—the depression was but a "temporary setback" and soon there would be lots of jobs and good times again. But the workers who had been laid off, put on part-time work, had their wages reduced, or who lived in fear of one or another of these eventualities, were already badly frightened and growing bitter with the conviction that they were the helpless victims of somebody's mismanagement. The closing of the wooden-heel factory, described earlier, gave them the first specific focus for the expression of their distrust of, and antagonism toward, the upper classes.

As jobs became scarcer in 1931 and 1932, workers and the general public alike began to urge that women and children should be eliminated from industry, as far as possible; jobs should be given first to men, to heads of families. The outcry against "child labor" resulted in a general movement to keep children in school longer. Among many of the newer ethnic groups, children formerly left school after finishing the eighth grade and went to work. Now there were no jobs for them, so many ethnic children went on to high school. The type of training they received there often prejudiced them against technological jobs as a satisfying means of gaining a livelihood. Many of these children began to want "white-collar" jobs instead. Others wanted to learn trades which would give them a higher social status than would unskilled factory work. Will Carlton's son, an upper-lower-class boy,[4] criticized the high school because: ". . . a fellow gets through there and can't go to college, has to work, he doesn't have a trade, has to start then and learn a trade. They ought to teach things like carpentering, electricity, something so they could go out and get a job."

Members of the upper classes became disturbed over such

4. *Ibid.,* pp. 185–188, for a profile of Will Carlton.

changes in attitudes, seeing in them threats to the established social hierarchy. Mr. Phillip Starr[5] (lower-upper) spoke for this point of view. In his mind he was lecturing the local school teachers. "Isn't it true that many of our boys expect to obtain a white-collar job when they leave school? Why is this so? Perhaps many of them never learn to develop an appreciation for shop labor or the manual arts because so much of their time is taken up with the study of books during the formative years. If they grow up to believe that practical things are not educational, we are defeating the very purposes we wish to serve. On the other hand, if boys learn at an early age that manual work is useful and honorable there is more likelihood that they will grow to manhood sufficiently appreciative of the occupations and of their real value to society."

The assumption implicit in these remarks was that members of the lower classes should be content to do unskilled technological work, that it is wrong to let young members of these classes acquire higher aspirations. The upper classes feared that, through the longer period of schooling enforced by the depression, members of the lower classes would encroach on upper-class prerogatives. Thus the upper classes felt the obverse side of the inter-class antagonism and conflict which bore most heavily on the lower classes because they were the most depredated by the depression.

In the competition for jobs, men of the lower classes developed attitudes toward women in factory jobs that were surprisingly similar to the ideas entertained by upper-class persons toward the thought that young members of the lower classes might come to compete with them for "white-collar" jobs. Sam Dixon's statement typified the male attitude: "Well, women shouldn't be holding jobs that men should have. All over the world everything is built on the home, the family, and when a woman goes out and gets a job she can't make the home, and the family breaks up. If women vote for women, women hold offices, what will happen ten or twenty years from now? There'll be a lot of women holding offices, and families will break up and then you'll be in a hell of a jam."

Sam was casting about, in these critical times, for any and all possible ways to bolster his fading sense of security. In times

---

5. *Ibid.*, pp. 141–152, for a profile of Mr. Starr and his family.

like these, he argued obliquely, women should stay out of men's shoes. A man had a hard enough time in maintaining his rightful status as provider for his family in competition with other men. Women should not make the going harder by competing in this field but should be content with their traditional place in the home.

And, in truth, working men were hard put to earn enough to enable them and their families to maintain their positions in the hierarchy of social classes; many had to accept the visible stigmata of moving into lower-class neighborhoods, were unable to provide "proper" clothing for wives and children, and even could not afford to eat the kinds of food they regarded as proper to their social positions. The great hardships of loss of social status were graphically told by Mrs. Timothy Pinkham, a lower-middle-class woman.[6] "A friend of mine is working in a shoe factory, getting six dollars a week. He is married, has two children, and has to pay four dollars a week rent. [In order to live in a neighborhood suitable to his lower-middle-class position in the social hierarchy.] That's terrible. The cheapest rents in the city are ten and twelve dollars a month, but for a respectable house anywhere that a person would want to live in, it would cost about seven dollars a week.

"Of course, you could go down on River Street and live down there, and rent would probably be cheaper, but who would want to bring up their children there? I guess it is all right for the Poles and some of the French who live down there. They seem to like it, but no one else would want to go down and bring up their children in that area."

For a lower-middle-class family to have to live on River Street was a tragic confession of failure to maintain social position, nearly as bad as being hungry. It was infinitely harder to bear because the worker lost his job, or had his pay cut tremendously, through no fault of his own. Under these conditions workers sought resentfully for someone or something to blame. More and more they came to blame the factory owners, particularly the absentee-owners, and the managers who had reduced their pay. These antagonisms were often generalized to include all members of the upper classes, who were not suffering the loss of status that lower-class workers were, who had plenty to eat

6. *Ibid.*, pp. 168–178, for profile.

and fine houses to live in, and who seemed not to care whether the workers were being harder and harder pressed.

The feeling of animosity toward the "rich" upper classes held by members of the lower classes at this time, as well as the desperate casting about by the latter group for some panacea to cure these social ills, was constantly being expressed.

Sam Jones (lower-lower), Riverbrooker,[7] expressed it well, in a conversation with Johnny the Greek, a lower-class shoe worker:

"Only the poor people help the poor, the rich never do. I know a man who works in the silver factory. He's got seven kids —yes and maybe eight, there's another one coming. He makes $28 a week and spends it all, for things for the kids. The other day he took in a mother and her little son and daughter and now he's helping support them.

"When anybody comes to see me hungry I give him something to eat. What would happen to that man if he went up on Hill Street? They'd say, 'Just a minute,' then they'd call a cop and get him arrested."

Johnny remarked: "The rich people drive like hell down the street. They are heavily insured, don't care how they drive. They run into somebody, run over a child, and they don't care. They go to court, get a smart lawyer, and get out of it. The poor people can't help themselves because of this."

Sam continued: "These poor guys are hungry; they're not only hungry for food but they're hungry for work. This is what happens when the rich don't spend their money. The poor do, they spend their money. I've lived sixty years, I expect to live fifteen more, and I expect it to get worse and worse as time goes on."

These accusations against the "rich" are of interest to us because it was by such constructs and episodes that members of the lower classes made concrete to themselves the chagrin they felt that some strata of the local society should be living in comfort while they were in danger of loss of status and self-respect. The lower classes felt that "in some way or other" the surplus wealth of the upper classes should be made available to them. Johnny, the Greek shoe worker, came close to the heart of this antipathy for the upper classes when he said: "Take all

7. *Ibid.*, pp. 168–178, for a profile.

the money in the world, if all this money was spent, kept turning over, why nobody would be poor, nobody would be out of work. The trouble is, these people like up on Hill Street, they have all the money, they keep it in the banks, they don't ever do anything with it and they don't spend it. If they spent it, why everything would be all right."

Sam Jones developed Johnny's thesis: "What they should do, they should make money every six months, and they should give it to everybody, and they'd say, 'Now at the end of six months this money is no good.' In that way people would always be spending money, there'd be lots of money, lots of work, everybody would have everything they needed because the rich people would have to spend money or it wouldn't be worth anything."

They came still closer to the realities of the situation when they recognized something of the effects of mechanization of industry on the working classes and proposed their idea of a remedy. Sam said: "A man has a million dollars and he gets a job to build a road, he makes another million dollars off the road. He takes these machines, steam-shovels, ditch-diggers, and the like, and he builds the road in a hurry. Not very many men work, they stand around on the outside and ask for jobs while he makes the money."

Johnny replied: "In times like this they shouldn't do that. They should take the machines and put them away and give everybody work, then when everybody had work, why they would get money and they would spend the money. Pretty soon there would be so much work they wouldn't have enough men. Then they could take the machines and bring them back again. They should save the machines until they don't have enough men, then they can use them."

There was the core of the economic argument that led the shoe workers to strike. They were convinced that the factory owners were making money and had jockeyed the operatives into a position that threatened starvation.

The notion was everywhere that depression was a definite, perhaps even a premeditated, attack by factory owners and managers on the workers. "If the working men," Sam said, "had treated the manufacturers in good times the way the manufacturers are treating workmen now, things certainly would have happened." Sam is saying what a shopkeeper said: "Working men are trustworthy and just; upper-class people are not." He is also saying that upper-class people can control what happens to our economy, that society is run as they ordain, that they force lower-class people to do their bidding. He went on: "The newspapers would have gotten up and howled about it if working men had taken advantage of good times but now they don't think anything about it." Here he implies that the newspapers are mouthpieces for the upper classes, that they give workers no hearing.

Will Carlton declared to an interviewer: "The owners cut the working man down every day. Some of these factories have given two 25 per cent cuts and the cotton mill, before it closed down, cut them down 45 per cent. I don't see why it closed down because I know they were getting cheaper labor here than anywhere else. [Even bankruptcy involves some obscure and discreditable conspiracy by owners and management.] Labor in the shoe factories here is cheaper than in Harrington or Baconsville or in any other place. [Our bosses are stupid or crooks.]

"That's been the trouble with the country: the rich people think the wages are too high. They've got to get them back low. The rich people don't want to see them handing out so much money for laborers, the working men running around in automobiles, so they're having this depression to get things back where they were. I don't think wages will ever be as high again. I don't see why the manufacturers care, though, they have plenty themselves."

Here is the open accusation that the whole depression was a conspiracy on the part of the manufacturers to attack the working classes. The conspiracy, he went on to charge, was not just general and nationwide, but also specific and local: "The Chamber of Commerce in Yankee City is strong for the manufacturers. They don't seem to want to help the town. They try and keep business out. Their slogan is low wages and open shops. They tried to get another silver factory in here some time ago but the Tooker Company was too strong in the Chamber of Commerce; they wouldn't let them in. The same way with the shoe factories; they don't want any more shoe factories as it would cause competition for labor. The manufacturers just run the Chamber of Commerce. They're just looking out for their own interests, not for the town."

Wages in the shoe industry had by this time decreased tremendously; workers suspected that part, perhaps even most, of the decrease was unwarranted, and that manufacturers were simply taking advantage of the hard times to force workers to accept low wages. Bessie Jones (wife of Sam), a shoe worker, said: "Wages are simply awful now. When a girl two years ago would be earning twenty-five a week, nowadays she is lucky if she gets ten. I know lots of girls who only get four. People are lucky if they get ten cents an hour. I guess they think we are overpaid if we get that.

"Some of the factories just cut down on the excuse of the depression. All the shoe factories are that way.

"If a girl is living with her family it is not so bad—she can get along. But when girls are alone, I don't see how they live. No girl can live honestly on four dollars a week. She can't go straight on four dollars a week." Here we encounter an accusation capable of rousing tremendous animosity of workers toward management and owners: "They are forcing our young women into immorality."

This accusation was leveled directly at one Jewish factory manager who did take advantage of the scarcity of jobs, it was said, by "forcing some of the young girls in his factory to go out with him," thus doing directly what other factories were accused of doing indirectly. This man's activities were exposed in the local paper, and general censure was heaped on his head. Will Carlton spoke the general opinion when he said: "It is these Jews and foreigners that run the shoe factories, and a girl can't get a job unless she goes out with the man. Things are awful over in the Big City Factory. That's about the worst thing possible, worse than starving. It's worse than anything to have a shoe factory run the way that one is."

Here the accusation is generalized against the ethnic (and non-resident) managers. It was assumed that native resident managers would not be guilty of such offenses, but these stranger-managers were capable of anything. Again, the increasing social distance between the operatives and absentee owners and managers of the Yankee City shoe factories was personified by the operatives. They attributed undesirable characteristics of the worst sort only to the stranger-managers.

It is clear that the depression in Yankee City intensified the class animosities shoe workers normally felt toward the upper classes and toward the managers of the factories as representatives of those classes. This antipathy for the whole of the upper strata of the local society was undoubtedly related to the envy workers felt for the people who enjoyed security while the operatives lost theirs. The antipathies and growing bitterness of the workers were directed particularly against management, for it was the factory manager who laid a man off or cut his pay. Workers convinced themselves that their rights were being purposely trampled underfoot by those at the top of the hierarchy. Charges with high emotional content were leveled at some of the managers which, added to the other wrongs workers believed themselves to be suffering at their hands, made the workers, by the winter of 1932–33, ripe for the open conflict which occurred the following spring.

2. Grievances

THE three major specific grievances of the operatives of all Yankee City shoe factories at the beginning of 1933 were: (1) that wage rates had been cut so much that operatives not only could not maintain a decent standard of living, but were unable to obtain even the bare necessities of life; (2) that the workers spent most of their time in the factories waiting for work, for which time they received no pay since most operatives were paid on a piece-work basis; and (3) that they were required to make, without extra pay, mostly novelty shoes which involved a great deal of extra work and militated against fast work because of the constant style changes. Let us examine these grievances one by one.

There is no doubt that wage rates had decreased very much by 1933. Various statistics compiled by the State of Massachusetts show that average shoe-operative earnings decreased between 45 and 50 per cent from 1929 to 1933, whereas the cost of living dropped only 23 per cent during approximately the same period. Average annual earnings of twenty shoemakers in one Yankee City factory showed a steady decline from $1,332 in 1930 to $924 in 1932. The average annual earnings of twenty-six women stitchers in the same plant fell from

$896 in 1930 to $581 in 1932. These workers were all full-time employees for the three-year period. There is no doubt, therefore, that shoe-operative incomes had had a greater proportional decrease than the cost of living.

The manufacturers were aware of this fact, but said that the decreases in wage rates were absolutely necessary because of the buyers' market for shoes. They pointed out that most of the shoes made in Yankee City were produced for the ABC chain of retail stores. The retail price in these stores had dropped, during the early years of the depression, from $5 per pair to $3.50. In turn, so the manufacturers said, the chain stated what they would pay for shoes; manufacturers either had to produce them at the price specified or lose the business to competitors. Since material and overhead costs remained fairly constant, the manufacturers pointed out, the only way shoes could be sold at the lower prices specified by the chain stores was to reduce either profits or labor costs. They had done both. One local producer said he had lost $25,000 in 1932, that it would have been much cheaper for him to liquidate his business than continue operations. But, he concluded, he felt he owed something to the working people of Yankee City in such times as these, and he did not want to add to the unemployment problem if he could avoid it.

Operatives felt, however, that the manufacturers were still making money, while the workers were taking the brunt of the depressed market in terms of pay cuts. They were further irritated, at this time, by additional mechanization in some of the plants. At the Jones and Jackson factory, for instance, two heeling machines were installed, in early 1933, and twelve men lost their jobs. The factory said the machines would make the heeling operation cheaper, but one worker said: "The heelers had been paid $2.34 and they said the machines would be cheaper. But I got the figures on the new operation and it was over $2.70 a case, including the royalty to the Shoe Machinery Company, as well as throwing twelve men out of work. I can't see how it was cheaper." This incident seemed to the workers to be a gratuitous cruelty on the part of the factory management; it injured the workers without benefiting the factory.

Regarding the operatives' second complaint, that they spent many hours without pay in the factories waiting for work, the manufacturers countered that they were spreading the work at the request of the President of the United States. By this device it would become possible to divide wages among many workers rather than discharge a large proportion of operatives.

But the trouble was that the manufacturers did not plan the work in a way which would give each individual two or three full days' work each week. Instead, they required all the employees to come to the factories every day and wait around for what work they could get.

To the operatives, this procedure amounted to working full time for part-time pay, and so there were indignant complaints that "girls work until six o'clock and on Saturday afternoons, and get from $7 to $13 a week." Or, again, "A girl assembler some weeks earns only $4, for which she works seventy hours. If she had only worked forty-eight hours, she'd have got less than a dollar." A girl operative said: "I made eight dollars last week, the most I've made in months. Some girls make only twenty-five to thirty cents a day, work all week and just get two or three dollars. We have to stay there all day—only work part-time but we have to be there when the stuff comes in. It isn't right. They shouldn't treat people that way just because they can."

These statements show that management and workers viewed the spreading of work from quite different and conflicting points of view.

Let us examine the third complaint voiced by all the operatives. Rapid style changes in women's shoes are a modern phenomenon, much accelerated during the depression in an effort to stimulate retail sales. In former years, a factory made shoes for half the year, building up stock in a local warehouse for the "Easter Season," then again for the "Fall Season." But by 1933 no manufacturer could make shoes for stock; style changes were too varied and rapid. This condition caused spasmodic factory operation—top speed on each order to get it out before cancellation, then a slack period of waiting for the next order. It had other evil effects, from the operatives' point of view. For one thing, a great deal of extra work, especially stitching and trimming, is required on novelty shoes. But the retail chain that bought most of Yankee City's shoes would not pay anything extra for this extra work because they sold all shoes at one retail price. Hence workers had to do this extra work without any increases in piece-rates—indeed, at reduced rates of pay during the depression. Moreover, workers had no chance to reach their maximum working speeds because the orders for individual styles were too small to give them sufficient practice. The bewildered shoemakers found themselves working harder than ever before, but earning much less.

The factory owners and managers said there was nothing they could do about this and referred again to the buyers' market. They argued that it was better for the operatives to have some work, even at reduced pay, than for the shoe factories to close down because of failure to meet the buyers' price.

Operatives were not satisfied with this argument. They felt they were being hoodwinked, that the owners were actually making money and taking advantage of the workers' dependence on their jobs to cut wages and make more money. They came to believe there was a conspiracy between the company controlling the ABC chain ("outsiders") and the local factories, especially the absentee-owned ones, to bleed the workers dry while they were waxing rich on the extra profits gained by racketeer methods.

Another source of worker discontent was the charge that Yankee City factories were paying differential rates for equivalent jobs and that some factories were paying the same piece-rates for fancy shoes that other factories paid for simple shoes which could be produced much faster. A worker said: "Land [manager of Weatherby and Pierce] is selling to the ABC Company the same as Jones and Jackson, and for the same work Land is paying less money. That's what people object to. One factory pays twenty-four cents a pair, another twenty-six cents, and another eighteen cents, all for the same class of work." The rates mentioned here were those paid to makers. And there is more to the complaint than immediately meets the eye. Land they regarded as a "capitalist" who had "sold out" to the Big City men, antagonistic to workers, seeking his own ends at the expense of his employees. Moreover, the eighteen-cent rate Land was paying each maker was being paid to each man on a three-man team, a total of fifty-four cents per pair of shoes for lasting. This was a greater cost to the factory than Jones and Jackson's rate of twenty-six cents a pair to each man of a two-man making team. But the workers ignored this fact to give vent to their feelings of distrust and dislike of the "capitalist." Although Land claimed his makers were working in three-man teams at their own request, the workers suspected some trickery somewhere, were sure they wanted two-man teams if Land wanted three-man teams, or vice versa, but could not decide what he really wanted.

The hazy sense of injustice cherished by Land's makers led them to open the hostilities between workers and management by going on strike in January 1933. Seventy-five lasters walked out, giving as their reasons the low piece-rate of eighteen cents and their objections to working in three-man teams. They had decided that three-man teams could not work faster and wanted to return to two-man teams at a piece-rate of twenty-four cents per pair per man. At the end of a two and one-half week strike the factory agreed to pay twenty cents per pair per man, not to reduce the rate for six months, and to permit either two- or three-man teams as the workers chose.

The operatives accepted these compromise terms because they "figured that in times like these they might get other workers," but they went back to work sullenly, still nursing a grievance. They said: "If the strike had lasted two weeks longer we would have gotten what we wanted, but we were poor and couldn't wait."

Operatives began to see that if they all banded together against all management and presented unified demands and a united front, they would have much greater bargaining power than had the seventy-five who had successfully struck against Land. This notion furnished the operatives food for thought as rates continued to be reduced through the winter of 1933, and the various abuses, as they saw them, continued.

Yankee City had had periodic strikes of shoe operatives in previous years, but there had never been a permanent union in that community as there had been since 1901 in the nearby town of Harrington (where the Shoe Workers' Protective Union started at that time). This union had made many efforts to organize a local in Yankee City. A former Yankee City shoe manufacturer said that, years before, while he was still in the shoe business, organizers had come down and talked to some of the workers in his plant. He had immediately offered raises and contracts to his employees because he thought unions were "a bad thing" and that it was desirable to keep them out of Yankee City. All his workers had accepted the raise, signed the contract, and declined to unionize—except for three men who had been with him a long time who wouldn't sign the contract, saying, "We've been with you a long while; you know you can trust us."

There had been another attempt to unionize Yankee City operatives in 1928. A union was organized in one factory and the workers struck because of a pay cut. The manufacturer hired strike breakers who slept in the plant and broke up the strike. "It finished the union, but the manufacturer went broke soon after because of the expense involved in breaking the strike."

An important reason why unions had never been successful in Yankee City was that the Riverbrookers, who were so numerous among the technological workers, would not join unions. In January 1933, a French-Canadian worker said: "Men from Riverbrook won't strike. They make fifteen dollars a week steady. They have a little house and garden. That's better than making forty or fifty dollars one week and nothing the next. It's always been that way here. A union never would work because of people like that who take just what they can get. They would rather have steady money all the time and not fight." But, finally, even the Riverbrookers began to listen to talk of unionization. Sam Jones, Riverbrooker, said, "We were always against the union but not this time; we're with the union now." [8]

Before the general strike started there were further rumblings and minor walkouts. On February 3, 1933, fifteen workers walked out of one factory "following some difficulty over wages." This strike, like the previous one in Weatherby and Pierce, was shortlived and unsatisfactory to the workers. The operatives were not yet able to present a solid front to the manufacturers, and these small groups of strikers were afraid

8. *Ibid.*, pp. 173–175, for a further statement from him and other Riverbrookers.

they would lose their jobs to some of the vast horde of unemployed workers if they persisted, alone, in their rebellion against management.

However, unionization was in the air. To workers whose pay had been greatly reduced, who lived in daily fear of losing their jobs, who were convinced that owners and management were taking advantage of the depression to grind them down while continuing to pile up profits for the factories, unionization came to represent the one hope of forcing the manufacturers to do "the right thing" for the workers. So the stage was set economically for the strike to come to Yankee City.

III

THE NATURAL HISTORY OF A STRIKE

1. Battle Strategy

THE industrial battle was fought between the owners of seven factories and 1,500 workers. Four of the factories, the larger ones, employed the vast majority of the workers and accounted for most of the $34,000 weekly payroll. This industrial war lasted a month. It began on a bleak and snowy day, March 10, and lasted until April 6. There were three clearly marked periods, each with different objectives and strategy; and in each the industrial workers and the managers were dominated by different feelings from those in the period preceding.

In the first period (March 10–19), when management and the unions fought desperately to gain control over the workers, the union was successful in organizing the workers, and management was prevented from regaining control. The second period (March 20–29) began when all the workers requested the unions to represent them in the struggle with management, and the unions, secure with the workers organized behind them, began frontal attacks on management. During this time each faction continued its intense efforts to influence and dominate public opinion in Yankee City. The union also won this fight since the public identified the union with the workers; and, as we shall show, most of Yankee City sided with these workers. The final phase (March 30–April 6), that of mediation and peace negotiations, began when the State Labor Board started a series of negotiations which terminated the strike. Other efforts at negotiation had been made from the beginning, but none was successful.

The ultimate objective of each side, to which each fashioned its strategy, was, of course, to make the other side capitulate and accept its demands. For management this meant that the workers would return to their benches under approximately the same working conditions and wages as those they had left.

For the workers it meant that management would agree to their demands and increase wages and improve working conditions. For the union officials it meant that the union would maintain its control over the workers and keep them members of the union organization, and that management would be forced to deal directly with the union rather than with the unorganized workers.

Each side organized itself and developed its strategies of offense and defense. The workers' defense tactics were centered around maintaining their unity and defeating management's offensive strategy of breaking up the workers' group and destroying their morale. Accordingly, the workers used ritual and ceremonial procedures composed of recognized symbols of solidarity, such as the flag, patriotic hymns, and the American Legion band. They achieved a defensive organization by means of meetings, speeches, entertainments, and the formation of a large number of committees which gave the mass of the workers opportunities to participate, to become and feel a part of a powerful and aggressive group. They took offensive action against management by making a series of demands for better wages and working conditions, by picketing, by attacks against management in the newspaper, and by use of the platform to influence public opinion.

Management's defense against labor was always to take the offense. The tactics management tried included sending foremen to talk to the workers individually, thereby separating them from the group, spreading discouraging rumors, advertising in the paper, insisting on secret balloting by the workers when they voted on the issue of returning to work, and, above all, threatening to move their factories elsewhere should the workers continue with their demands and join the union. Of course, it must be remembered that each side, throughout the strike, was being deprived of its income—labor its wages, and management its profits.

Before recounting the progress of the strike it is necessary to give further consideration to the character of the workers and of management. Half the managers were local, most of them Yankees. Mr. Land, local man and spokesman for the owners, Mr. Jones, Mr. Jackson, and Mr. Pierce were born and reared in Yankee City.[1] Most of the outsiders were Jewish. Mr. Cohen and Mr. Shulberg, who controlled three large factories, were from New York. Mr. Bronstein and Mr. Luntski were from Boston.

The union had a very difficult problem in maintaining unity among the hundreds of workers because of the heterogeneity of the group. Over 40 per cent were women.[2] The median age of the workers was thirty-five years for the women and forty years for the men.[3]

The workers were scattered through the three lower classes, and these status differences added to the difficulties of forming a cohesive group: 10 per cent were in the lower-middle class, 44 per cent in the upper-lower, and 46 per cent in the lower-lower class (see charts 12 and 13).

2. The Union Fights Management for Control of the Workers in the Community

It is not definitely known whether Yankee City shoe operatives asked to have union organizers sent over from the nearby town of Harrington, or whether the union took the initiative. According to one report, the Yankee City shoe workers sent a delegation to request the Shoe Workers' Protective Union officials, who were conducting a strike in Harrington, to come to Yankee City and organize the operatives. On the other hand, a forewoman in one of the Yankee City factories said she had warned her employer two weeks before the strike that the operatives would walk out on March 11. She said she knew about it because some of her Harrington friends who were in the union had told her that the strike was coming to Yankee City next.

At any rate, on the morning of March 10, about three hundred of the striking Harrington shoe operatives came to Yankee City and marched about in front of the factories shouting for

---

1. See Chapter VIII of this volume for a full-length portrait of most of these men.
2. We are using a sample here of 985 workers.
3. The social characteristics of the workers are treated in a detailed appendix (6), "Social Characteristics of the Shoe Workers." Tables 14 and 15 give a complete comparative summary of the age, sex, ethnic group, and place of residence for the whole group.

the Yankee City workers to join them in a protest strike. Here was the first public manifestation of shoe workers joining forces across community lines. None of the local workers walked out on this day, however, but they watched the parade from the factory windows.

Early in the evening of March 10, the president of the Shoe Workers' Protective Union, Nixon, and other union organizers came over from Harrington to talk to individual workers. Later, one hundred and fifty Yankee City workers went to Harrington to hear Nixon speak to the strikers there. On Saturday morning, March 11, the union officials were again in the city. Headquarters were established in the Yankee City Hotel building. The S.W.P.U. called a mass meeting of all shoe workers for Monday evening and proposed a city-wide strike such as that in Harrington and elsewhere. As a result of this preliminary agitation, workmen in some of the factories left their jobs on Saturday.

On Monday morning, March 13, a meeting was held in the Yankee City Hotel building. The cutters of Bronstein's started a parade past the principal factories; hundreds of workers walked away from their jobs and joined the parade. It was estimated that three hundred marched from Bronstein's. They were joined by practically the entire force of Jones and Jackson, some five hundred workers. They marched past Weatherby and Pierce, but no one came out, then past Luntski's and on to the City Hall where a mass meeting of more than twelve hundred persons was held. The organizer of the SWPU in Harrington headed the march and conducted the meeting. Enthusiasm was great; the workers were in high spirits; some sang. Union officials, to create an emotional bond of unity among the workers as well as a favorable emotional attitude toward the union, gave the women American flags to wave. The mayor had granted the strikers free use of the City Hall for their meeting. The announcement was made that the workers did not intend to go back to work until they were granted "decent pay." At the meeting, various crews in the different factories selected committees to represent them in the strike. Here, again, the union officials were using their influence to create a social structure out of the amorphous mass of strikers, preparatory to the attempt to persuade them to join the union.

It was explained at the morning meeting by one organizer that the W. and P. employees would strike after finishing their shoes for the day. By mid-afternoon they were out. One of the union organizers marveled at the way all the workers joined in. They had expected to spend a week on the organizing before workers would come out.

On Monday night, the general president of the union addressed mass meetings at both the City Hall and Pythian Hall.[4] He made frequent reference to the "New Deal" and other such phrases which had become by-words since the change of the federal administration. He exhorted the workers to remain intact and to keep solidly behind their committees, which were striving hard to bring about better working conditions in the city. He was here suggesting an identification of the workers' committees with unionization, further inclining the workers toward joining the organization he represented. He explained in detail the success the organizers had met in efforts to organize in other cities. He said that a general wage increase would be demanded. Other union officials, including an organizer formerly of Yankee City, spoke encouragingly to the strikers. Union cards were distributed, but details of signing up were left until after the demands would be drawn up for presentation to the manufacturers. The officials were using good tactics on the workers—tacitly assuming they would join the union but seeming not to force the issue—"you pay no money till you see some results."

On Tuesday the Yankee City daily paper reported:

---

4. Associations like the Knights of Pythias, the Ancient Order of Hibernians (which also loaned its hall to strikers), the Elks, Knights of Columbus, etc., whose memberships derived largely from the three lower classes, favored the strikers from the start. Other associations whose memberships were predominantly upper-class—the Rotarians, the Dolphin Club, etc.—took the employers' side throughout the strike. The Chamber of Commerce divided its allegiance: the members who owned small retail stores were mostly lower-class and obtained their trade from the workers—they were for the strikers; the members who were manufacturers or owned large retail or upper-class stores were against the strikers. The Chamber of Commerce played an important role in the strike, and this split in its allegiance became important in later phases. The response of these various organizations to the strike situation furnished a revealing insight into the way the shoe operatives were integrated into the larger social life of Yankee City. It was an instance of the "broad base" integration of the factory in the community.

The city's main industry, the shoe business, today is at a standstill. All the shops have closed their doors. . . . It is said that not a single shoe is being turned out in the city today. The shoe workers presented a united front last night at their first mass meeting. City Hall auditorium was crowded to overflowing and Pythian Hall on Main Street was the scene of another gathering of capacity size. Great enthusiasm was manifest at both places among the great crowd of men and women and demands of a new wage scale, which is being drawn up, will be made to manufacturers immediately. . . . Everything so far has been orderly and there is a happy spirit noticeable among the men and women who have left their benches seeking better wages.

It will be instructive to follow the attitude of the *Herald*, the principal medium of communication in the city, during the course of the strike. Each side used the *Herald* and tried to control it. The paper realized that the small merchants, who were the source of much of its advertising income, were largely on the side of the strikers; therefore, it could not take a position against the workers without running the risk of losing revenue. At the same time, the paper feared the union and also feared the effects of the removal of the absentee-controlled factories. Hence, it tried to play both sides of the conflict, sometimes swinging one way, sometimes the other. We shall remark on these swings from time to time. The note at first was one of happy optimism, though strictly non-committal on the issues involved.

Tuesday morning, March 14, saw another parade of the strikers, numbering more than fifteen hundred, headed by the American Legion band and accompanied by a fife-and-drum corps from one of the factories. Many of the marchers carried signs demanding better wages. They marched about five miles through streets slushy with snow and rain. Girls and women were among the marchers. The parade went past all the factories that up to a few days ago had been busy but were now closed because of lack of workers. The parade and the band were further emotional appeals engineered by the union officials to convince the workers of their solidarity and the desirability of unionization.

Merchants contributed largely to the fund for hiring the

American Legion band. These small merchants were sympathetic with the strikers for two reasons: (1) they, too, were lower or lower-middle class people, interested in the welfare of others of comparable social status; and (2) they sold their merchandise almost entirely to shoe workers and other members of the working classes and had to show sympathy for their cause to keep their trade. At a mass meeting of strikers it was announced that the price committee would furnish lists of the rates at which they had been paid, in order to furnish the organizers with a basis for demands to the factory owners. Again the keynote was: "a general wage increase will be demanded."

On Wednesday, March 15, one of the factories tried to open, and twenty-eight men went to work. The union leaders told strikers to go in and get a job, spend the day working, and get the names of all those who were working and then try to talk them out of it. The strikers went in, spent the day, and the next day no one showed up. The place stayed closed thereafter.

At a mass meeting of more than a thousand strikers in City Hall on Wednesday morning, Nixon, the union president, and Hughes, of Harrington, the general secretary and treasurer, were the speakers. Both expressed confidence that the strike would be over within a short time and that the workers would be victorious. Nixon stated the union objects: to do good for the workers and to stabilize the industry so that the employers could do right by their workers and also make money. He implied that poor management had caused the chaotic conditions existing in the shoe industry, that the union could do, for both management and the worker, what management itself had failed to do. He said that the number of working days should be cut from six to four in order to give the workers three days' rest. He said further that it was not proper to battle labor-saving machinery, but that workers should control the machines and put them to work for them.

The morale of the strikers was high. It was announced that already six hundred had signed union cards. The union was rapidly achieving its first goal: to get control of the workers. The price committee met in the afternoon to draw up demands on the factory owners.

An organizer from Harrington stated he was greatly pleased with the coöperation shown by the police department, merchants, the mayor, and others. This statement was calculated further to convince the workers, as well as the general public, of the respectability and acceptability of the union. The good will of the community had to be captured to assure success. Some of the strike leaders had visited the mayor, who offered his services as chief executive to effect a settlement between the contending parties. The strike leaders declined, with thanks, saying that there was no need at present for his services. The mayor wanted to demonstrate his power and political control over Yankee City affairs; the union officials wanted to gain undisputed control over the workers.

In Thursday's paper there appeared a letter signed by Mr. Land, owner-manager of Weatherby and Pierce, which was the first indication from the manufacturers that *they admitted there was a strike*. It is interesting to note that Land, a local man, was made the spokesman for the manufacturers in the interchange of open letters to the *Herald*. The absentee managers and members of the absentee-controlled hierarchy, in general, stayed in the background, although they had more at stake in terms of volume of production in Yankee City than did the locally owned and operated factories. There is little doubt that the manufacturers recognized the in-group prejudices of shoe operatives and the Yankee City population in general and they selected a local "front man" to take what advantage they could of the favorable feelings toward him as a local citizen. Except for this concession, however, they assumed that public opinion would be with them and against the strikers from the start. The opening letter by Land shows this assumption in its sarcastic references to the workers which he thought would make them appear ridiculous to the citizens of Yankee City. After the manufacturers had been censured on several counts by public opinion and it became evident that the complaints of the workers were made seriously, the owners were bewildered and resentful.

In his letter Mr. Land said:

It's a sad thing to see the shoe workers of Yankee City, intelligent and far above average in well-being, allow themselves to be called out from their work and misled as to actual conditions prevailing in the shoe business. Falling prices are not confined to the shoe business. It started with commodities, which are the farmers' wages. We don't sell our shoes to shoemakers. There are many more farmers and their dependents than shoemakers in this country.

Somebody has come in here from the outside and, with promises that they cannot keep, has drawn men and women away from their daily labor and the money that they would earn, and has already begun to exact payments from them.

Yankee City has had less shrinkage in payrolls and there has been more money per capita paid out by industry to the workers of the city of Yankee City than any other city in the commonwealth. I think this is why savings bank deposits are highest and this is why welfare cost per capita is less than any other city in the commonwealth, and for no other reason. It's the logical answer, isn't it? We urge the cleaning up of bad conditions, bad jobs, short change, unfair practices, where they can be shown as such. There are plenty of agencies established by the state for such emergencies and we don't need to support a third party at any time to accomplish the desired results.

Seventy-five per cent of the workers that are now out don't know what they are out for. Seventy-five per cent of them didn't want to leave their work at all. Fear, intimidation, the thing our president says should be banished, is responsible for their being out where they are.

Management doesn't blame the workers for the conditions of markets. Who can tell us of some successful managements that have made money in the last year? The banking holidays are evidence that management of all business is unable to cope with the situation. It's something beyond anybody's control. It's the result of a World War and bigger than this county or Massachusetts or the wages shoe workers get.

We stand ready to prove that the wages we pay are fair and if they are not, adjust them, and we will do it through the state board of arbitration, which is our legal right.

Meanwhile, a meeting of the manufacturers to which Mayor Mayfield was called was held on Thursday at Bronstein's.

Cohen, head of the ABC Company, was there. The mayor was asked to go back to City Hall and into the meeting of the strikers and demand that they return to work. The mayor said that he had no authority to do this but that he would try to see Nixon and get him to arrange for representatives of the strikers to meet with representatives of the main factories. The manufacturers here attempted unsuccessfully to get the mayor to coöperate with them in refusing to recognize the union. They had assumed that the mayor, as well as the public, would side with them against the workers. The mayor, after all, belonged to the upper classes and was as important a figure in the community as were the factory heads; they belonged to the same associations, and so on. The manufacturers were shocked when Mayfield insisted on remaining impartial. They accused him of siding with the workers, of deserting his own social class. Mayfield felt he remained neutral throughout. This was the impression, too, of the great majority of his fellow citizens and of impartial onlookers. This meeting was not reported in the *Herald.*

On Friday morning, at a meeting of fourteen hundred striking shoe workers, Nixon announced that nothing had been accomplished in the way of negotiations with manufacturers, although he had talked with officials of the major factories. At Bronstein's he had been told by the superintendent that any settlement would have to be made with the ABC concern. The manager of Bronstein's was a minor figure in the expanded hierarchy and could not speak for himself as Land could. Nixon was informed by telephone that Weatherby and Pierce was "not interested." Cohen, of the ABC Company, finally said he was willing to meet the general president. At one factory there was "some talk of an agreement, and the owner promised that when a settlement was made he would donate $500 to the union to carry on its work." Nixon was to confer with officials of Jones and Jackson on the next day. He again stated the union's objectives at the strikers' meeting: to stop wage cuts; to establish union closed shops; and to have standard lists of wages. This was the first mention of a closed shop as one of the objectives in the Yankee City strike. Nixon was striving to force the manufacturers to confer with him and recognize the union. At the same time, he fought to maintain control of the workers by holding out rosy promises of what he could do for them. There was more entertainment at the meeting: singing and dancing by some of the strikers. In this, union officials showed themselves to be sound practical sociologists in their leadership of workers. Their goal was to weld the workers into a cohesive social group, the union, and to pit them against all manufacturers. The solidarity was achieved through the various "secular rituals" we have mentioned: speeches by which workers' sentiments were reinterpreted to them in union terms; parties, entertainments, flag waving, bands, and "can days," [5] all designed to produce a festive spirit of comradeship; committees, picketing, and parades to make the workers feel they were doing something to accomplish their ends. By all these forms of participation, union officials influenced workers to *feel* their unity and their opposition toward all owners.

Sam Dixon, Johnny the Greek, and several other shoe workers from the Jones and Jackson and the Weatherby and Pierce factories composed a letter. It was said that someone from the union helped them say what they had to say. The letter appeared in the paper as a reply to their boss:

Mr. Land talks about paying fair wages to his help. In looking over the wage lists of other firms in the city, we found Mr. Land well down on the list, in fact, close to the bottom. We only used price lists of shops making the same class of shoes, and other factories out of the city are making the same shoe with better conditions and wages twice as much as Mr. Land is paying.

There can be little doubt that these arguments were learned from the union officials. The union's fight is half won and the manufacturers' half lost when workers accept union teachings as truthful and reject the claims of manufacturers.

How about a room girl making $8 per week! Can a girl remain clean and pay her way on such small money? What about the state laws? We were not misled. We had to call in outside help to protect the workers of Yankee City.

5. Days set apart by the union for contributions to alleviate distress among families of striking workers.

Mr. Land urges the cleaning up of bad conditions. That is what we are going to do.

The conference between the Workers' Emergency Committee and eight men representing the manufacturers was held that afternoon. Shulberg and Cohen of the ABC Company were present as representatives, contrary to the mayor's advice. In advising against their presence at this meeting, the mayor was recognizing the social distance between them (men near the top of the absentee-controlled business hierarchy) and the local workers. The mayor knew the Big City men's presence would antagonize the workers, who were suspicious of them because they were both strangers and Jews. No progress was made at the meeting. The manufacturers informed the strikers' committee that they were not willing to deal with a union.

Thus ended the first period of the strike: one of activity and organization by the strikers, of parades, and mass and committee meetings. No direct contact had as yet been made between the strikers and the manufacturers. But the union had won the first campaign because it had obtained and kept control of the workers in spite of the manufacturers' attempts to prevent it.

3. *The Union Takes Control and Attacks Management*

THE second period began with a meeting of the strikers on Sunday, when the committee presented its report. The strikers unanimously voted to have the union president represent them in all negotiations. The union having fought and won its first battle, the fears of the manufacturers had been realized: the union had definitely acquired control of the workers.[6] Nixon told the workers to have faith in their cause and their leaders, and they would win. He stressed the necessity of presenting a united front and said that only through sticking together could they win. He charged that seeds of dissension were being sown among them by emissaries of the employers. (Foremen were said to have been sent out by their employers to interview workers with the idea of persuading them to return to work.)

He said further that he had been informed that one of the local manufacturers had invited his workers to meet him in his

---

6. The union in 1945 belonged to the C.I.O.

factory to talk over the situation, adding that if there were such a meeting, it would be in City Hall or A.O.H. Hall with the union representatives, not in the factory. Nixon was here seeking to insure the maintenance of union control over the workers. He also assured them that the manufacturers were already weakening. He spoke of the necessity of sticking to the demands—as drawn up by the S.W.P.U.—which follow: (1) closed shop; (2) 10 per cent increase in pay and standard rates for all factories; (3) a forty-eight hour week except for three summer months (forty-five hours); (4) the latter two agreements to be effective until July 1, to be then re-opened for discussion if desired by either side in a local board of arbitration.

Union officials announced at the Sunday meeting that they had heard that the shops would be open on Monday and would run if enough employees returned. A call was made for volunteers to do picket duty. Five hundred responded. On Monday morning, all the factories were patrolled by pickets, even though Luntski's seemed ready and willing to sign an agreement with the union. Here we see an important characteristic of the institutionalization of conflicts between horizontal levels —the strike now involved actively all workers as union members against all manufacturers. Individual differential agreements had become impossible.

Jones and Jackson requested police protection. Six officers were detailed for duty, but there was no trouble. The picketers knew the officers and there was friendly exchange of repartee. No guns or night sticks were in evidence. Here the absentee-owned factory had attempted to subordinate the strikers by calling in the symbol of political authority. The attempt failed because the workers and the police were both members of the lower classes in Yankee City society and were personal friends. The sympathies of the police were with their friends, the strikers, not with the absentee owners of the factory.

From Monday, March 20, until April 6, the end of the strike, all the factories were picketed. Volunteers were called for, their names taken, and they were assigned to duty. Pickets were changed each hour, those in the picket line being taken to and from factories in automobiles. In order to avoid possible future discrimination on the part of employers, workers were not asked to picket their own factories. This was also intended to develop group feeling and total responsibility rather than individual relationships: i.e., shoe workers as a unified group against factory owners in general and not shoe worker in particular plant against factory owner in particular plant. Women and girls participated despite the cold and slushy weather. The picketing was always orderly and quiet. This was a bid by the union and workers, of course, for public support and sympathy. Pickets were given hot coffee and doughnuts by local storekeepers. By this means the shopkeepers showed their sympathy for the strikers and strove to protect their own business interests at the same time.

On Tuesday, at a strikers' meeting, a strike leader from Harrington, in charge of picketing there, complimented the picket line of the day before. He said that one hundred and fifty men had come down from Harrington and volunteered to do picket duty. Enough local men were doing the work already, however, so the Harrington men were not needed. Entertainment, as usual, was furnished at the meeting after the speaking; there was music by an orchestra, singing, acrobatic dancing, a comedy act, a harmonica selection, and selections by a mandolin group. Nixon reported his conversation with two of the manufacturers and said he expected to have another conference with them the next day. The manufacturer solidarity was apparently cracking somewhat, some of the managers acting independently of their group. The *Herald* reported on this day that "all but a few of the two thousand workers are said to have joined the Shoe Workers' Protective Union."

The word was passed on Wednesday, March 22, that Bronstein's was to give its workers a party at the factory at two o'clock that afternoon, the party to be accompanied by a talk. As a result, the number of pickets was increased to several hundred, and for several hours they patrolled the whole block in which the factory was situated. Two policemen were on duty at the door. Nobody entered the factory and the party that was scheduled could not be held. Once again the union had successfully fought off an attempt by a manufacturer to deal directly with the workers.

At the mass meeting on Thursday, Southwick, of Harrington, organizer of the strike at the Ratsey Shoe Company in Saltville, asked local strikers to go to Saltville and take part in a parade there to be led by one of the town selectmen. Many persons volunteered to parade and did so, all workers conducting themselves in an orderly manner. At this meeting, also, Nixon praised the strikers for having prevented the party by which Bronstein's had hoped to lure its workers back to their benches. All this was calculated to increase worker solidarity still further, not only among Yankee City strikers themselves, but also between them and similar workers in other communities.

The local paper continued to serve as the battleground on which these two groups fought for public support. We have already quoted letters published before the workers accepted the union. Those we shall give now were written after the operatives had joined the union but before the manufacturers were willing to accept it and deal with it. It is significant that the union officials themselves took no active part in this battle; letters representing the union point of view were signed by workers or groups of workers. The union officials appeared, all through the strike, to be better tacticians than the owner-managers.

Mr. Land ("spokesman for us," said the owners; "that lousy Kike-lover," said the workers) addressed his help through the public press:

My friends, you are trying to lift yourselves by your boot straps.

One of our customers in Boston cut his price last week to $3 a pair. Another one cut his from $5 to $4 on hand-turned shoes made in Brooklyn.

The last trip two weeks ago made to Chicago to see a mail-order company, who have been retailing some of our shoes for $5 a pair, they notified us that they were not going to consider any shoes at more than $3 for the coming season. That simply automatically lets us out.

Our normal sales are 20,000 pairs a month. We have sold 17,000 in the last three months and none in the last two weeks.

Sam Dixon and his friends picked up their boss' theme in a reply:

This strike is not only local but in all cities where shoes are manufactured. It will soon be universal. Why? Starvation wages and deplorable "Buyers" are in the saddle, thanks to illegitimate and racketeer manufacturers. We aim to remedy this condition, and protect the manufacturer with our organized support. The manufacturers have not the intestinal fortitude to protect themselves so we have decided to protect ourselves as well as our employers.

No right thinking man can object to the union and it is a fact that reputable firms welcome it.

They then used "the punch" which they knew would hurt the boss most of all:

We have at the present time three manufacturers in this city making shoes for the same buyer. One of these concerns has always maintained a model factory and was an asset to the city. The wages were paid dollar for dollar with the best factories in this vicinity but they cannot pay them any longer, they are whipped into line and to compete with the other two factories they will be forced to adopt and impose the conditions mentioned in the foregoing paragraph.

This is a reference to Tim Jones who, partly because he was a native of Yankee City, had always been regarded as the workers' friend. Even he, the workers are pointing out, is unable to resist "racketeering" methods.

On Monday, March 27, Land and Sam Dixon's crowd exchanged verbal blows. Land pleaded with the people of Yankee City to "save the industry and life of Yankee City. A workable plan has to be formulated," he said, "to put the shoe business and every other Yankee City industry in a safe and sound position."

Sam Dixon (and his group) replied:

Mr. Land, everyone knows that to end the depression we must increase wages and put men to work. Someone has to start and we, the shoe workers, are the pioneers. Other industries will follow and we will get back to normal. You admit that the shoe manufacturers have been anemic. Well, I have seen anemic persons who possess what the shoe manufacturers haven't shown in regard to the marketing of their shoes; and that is "spunk." Haven't you a National Boot and Shoe Association? What are they doing? Are they writing advertisements for the newspapers whining and crying about their nice help and their nice shoes? Evidently, yes. Now, Mr. Land, show your spirit by being the first to sign up with the union. You are a leader—the others will follow. Show the same effort with us; we will be with you 100 per cent.

Bronstein says he thinks so much of our city and you [the Bronstein employees]. Ask him why he doesn't live here, as all our former manufacturers did.

Ask him if he has a lease on his factory. If he has, for how long. The public desires to know our side—tell them. They ask you to coöperate. You know, shoe workers, you are the "co" and they "operate." Ask him what he means by spending $120,000. Do the ciphers represent your wages?

On Friday the strikers presented their demands to the factories. Nixon, addressing a meeting of some thirteen hundred strikers at City Hall that day, said it appeared as if they would win. The manufacturers were "on the run" and would have to promise higher pay and different working conditions soon. He interpreted the manufacturers' mere recognition that there was a union as the beginning of capitulation.

In a spirit of reasonableness the owners publicly stated that "low wages were universal" and promised that "when there is an increase in shoe prices the raise will be reflected in the wage scale."

On Saturday, March 25, Commissioner Wadsworth of the U.S. Department of Labor drew up a recommendation for the strikers and manufacturers addressed jointly to Land, as chairman of the committee of shoe manufacturers; to Nixon, as president of the union; and to members of a committee representing the striking employees. It did not mention the union.

The manufacturers signified their intentions of signing. On Sunday, the 25th, the striking shoe workers (almost thirteen hundred) met at City Hall and unanimously voted against the recommendation of the federal conciliator because it did not recognize the union. According to the mayor's (later) comment, "they (the federal conciliators) couldn't do anything."

It was too early in the day." Nixon said: "Public authorities are supposed to keep a neutral attitude, but obviously they didn't show a neutral attitude" (i.e., not recommending recognition of the union).

So matters were again at a deadlock. The strikers were exhorted by union officials to hold fast and remain courageous because the factory owners would soon come to terms. Extra pickets were put on. Announcement was made of a "can day," when donations would be received and the proceeds given to families of shoe workers in distress.

On Monday, March 27, at the instigation of the chairman of the Industrial Committee of the Chamber of Commerce, a "representative group of the business interests" were told the manufacturers' side of the controversy at a meeting in the office of Bronstein's. All important business interests of the city were represented in this group. At the meeting, Cohen, the president of the company, invited to Yankee City by the factory heads who sold shoes to the ABC chain, is reported to have stated that the ABC Company would be obliged to accept the offers made by other cities to manufacture shoes if a union were insisted on in Yankee City and that ABC Company factories would exist in other cities under union conditions but not in Yankee City. Here was a clear expression of independence from community control. The president of the chain, whose main offices were in New York, could easily carry out the threat.

On Wednesday morning, the 29th of March, after meeting with the strikers' committee, the merchants' committee met with the shoe manufacturers and read the proposed agreement of the strikers. The manufacturers objected to union recognition and decided to draw up an agreement of their own. The committee of operatives delivered the ultimatum: they would not agree to any proposition proffered by the manufacturers that did not carry with it the signing of the regular agreement with Local Thirty-Nine of the Shoe Workers' Protective Union. This ended the meeting, leaving the situation deadlocked.

Thomas Brown, leader and opinion-maker (quoted earlier), summarized the position of his group when he said publicly:

"The strike of the shoe workers in this city has arrived at such a serious stage that it is threatening the very industrial life of Yankee City.

"My interests are wholly from the standpoint of the welfare of Yankee City. I am convinced from communicating with the people who control the output of three of the factories in this section and contribute seventy-five per cent of the total weekly payroll that they will never sign articles which recognize union control and operation. Whether we like it or not it seems to be a fact. It also appears quite certain that, if both the employers and the employees remain firm in their determination to stand pat, at least three of our largest factories will be lost to Yankee City, which will cripple it industrially for many years to come, for there is not the remotest chance for new shoe factories or any other industry locating here for many months to come." This was the grave fear that inspired his words and gave many business men sleepless nights. The three companies that threatened to move rather than submit to unionization were the ones that made shoes for the ABC chain. Their refusal to accept the union was dictated by the president of the chain; neither the local managers nor the community had any control over this individual.

Mr. Brown continued:

"We have already lost fourteen or fifteen manufacturing concerns since the depression started. We cannot afford to lose more. It must be borne in mind by all that no manufacturing concern is able to make a profit at the present time. It also must be taken into consideration that those factories which are on the verge of leaving employ many workers whose homes and families and all their property and interests are in this city, and if thrown out of work permanently will cause a real problem both to them and to Yankee City. This condition must be avoided, if possible.

"For these reasons it is necessary for some serious thinking on both sides before it is decided to wreck the city socially and industrially.

"Wholly with the welfare of Yankee City and its business and industrial interests in mind, I make the following suggestions as a basis of settlement:

"First, that the workers be allowed to retain their union organization and that the manufacturers negotiate with representatives of this organization, either as individual firms or collectively as may be decided, but that the workers do not insist on the signing of articles of recognition. The union organization can be retained then as a safety valve for adjusting further grievances pertaining to either prices or working conditions which may become unsatisfactory in the future.

"Secondly, on the part of the manufacturers, they agree to the immediate arbitrations of price schedules and working conditions, so that the workers may return to their benches without further loss."

During the course of a citizens' meeting, Mr. Kirk of the Citizens' Committee suggested polling the strikers by a secret ballot as to whether or not they wished to return to work. Jackson, of the firm of Jones and Jackson, made a plea that the strike be settled at once, saying that the ABC Company would be obliged to move from Yankee City if it were not. Rafferty, a former mayor and reputedly an ardent advocate of the people's cause, asked why, if the union could be recognized by the ABC Company in Bayville, it couldn't be recognized in Yankee City. Nixon followed with an argument calling for the ending of hostilities. He said that at least two Yankee City manufacturers did business with labor groups in other shoe centers. He believed that dealing with the union would be beneficial to the industry. In regard to the secret ballot, he said he would not intervene if it were wanted by the workers. Nixon apparently was worried over the length of the strike. He did not insist on such open control over the workers as he formerly had. The strikers voted against the secret ballot. The manufacturers' agreement was voted down.

4. *Peace Negotiations—The Third Period*

THE State Board, which had been notified of the strike on March 20, had not sent mediators to Yankee City by the 29th, so the mayor went to Boston to enlist their help. It reported that its staff had been busy with other affairs, but Mayfield thought they had stayed away purposely until the strike got fairly well along with the idea that they would be more apt to get an agreement.

The State Board of Conciliation and Arbitration consisted of three men appointed by the governor to serve three years, one representing labor, one the employer, and a third, the chairman, representing the public. The Board had no coercive powers, but could hold public hearings, subpoena witnesses, and obtain complete records of any industrial dispute. After deliberation they usually published their findings, fixing the blame for the strike, and relied upon public opinion to settle the matter.

The three members of the State Board came to Yankee City on Thursday morning, March 30, and met with the strikers' committee and the manufacturers in a private hearing to acquaint themselves with the stands taken by the opposing sides. The strikers (about twelve hundred) gathered in City Hall in the hope that the Board would have a public hearing. The Board's recommendations were given to the *Herald* by Nixon on Friday, the 31st: recognition of the union (with certain reservations and interpretations), the State Board to be the arbitrating body; the agreement to be in force until January 1, 1934; and employees to return to work immediately without discrimination.

On the same day an article appeared in the *Herald* headed, "Merchants hard hit by the shoe strike . . . One milk dealer estimated a loss of business of twenty dollars a day. Retail business is practically at a standstill. Stores usually well filled on Fridays are practically deserted. Unemployment has cost the storekeepers thousands of dollars." The paper was now doing all it could to effect immediate settlement of the strike, on any terms. By stressing the losses of small retailers in this article it avoided offending them, and even hoped to enlist their active pressure on the strikers to take whatever terms they could get and conclude the strike.

On Saturday, the manufacturers met to deliberate on the recommendations of the State Board. They drew up a new agreement which was presented to the strikers at a mass meeting on Sunday. It included the following provisions: (1) union recognition; (2) arbitration by the State Board; (3) no pay adjustments until after August; (4) agreement to remain in force until March 1, 1934; (5) employees to return to work immediately.

The workers on Sunday rejected the offer of the manufacturers and demanded a 10 per cent increase in wages before going back to work.

On Monday, the State Board again conferred with the strikers. The workers were willing to enter an arbitration agreement provided that any pay increase would be retroactive to a definite date.

On Tuesday, April 4, the strikers' committee met with the manufacturers. Both sides had agreed on fifteen out of sixteen points of difference. The exception was the question of when the wage increase was to be effective. Following the meeting the manufacturers issued a statement withdrawing all former propositions and announcing the opening of factories to all strikers who wished to return. The *Herald* reported, "Settlement appears more remote than at any time since the walkout."

On Thursday, April 5, the State Board met with the manufacturers. Notice was given that it might be necessary to call a public hearing to place the responsibility, then leave the next move to be forced by public opinion. As a result, two shoe firms, Luntski's and Weatherby and Pierce, offered to sign contracts with the union. The workers refused with the announcement that all the factories must sign.

On Friday, April 6, the State Board met with the executive committee of the strikers, and on Saturday morning the strike was settled. The manufacturers accepted the union, and the strikers accepted arbitration by the State Board on the matter of wages. The workers were to go back to work on Monday at the same wages which prevailed when they went out on strike. The State Board was to make an investigation of the wages and announce a decision to take effect at once, but not to be retroactive.

Thus ended the shoe strike in Yankee City. It had lasted a month, had tied up the entire Yankee City shoe industry, and had cost it many thousands of dollars. The workers were jubilant, confident that they had won a great victory. The mayor complimented the president of the union. Nixon complimented the workers on their attitude throughout the strike and praised the executive committee. The manufacturers felt that in some measure they had won a victory or at least they had not lost everything because, although they had to agree to recognition of the union, they had persuaded the State Board to arbitrate the wage rates, which would relieve the manufacturers of a good deal of unpleasant work, and they had been able to prevent the making of any wage increases retroactive.

Plainly, labor had won its first strike in Yankee City, and, even more plainly, an industrial union had invaded the city for the first time and had become the recognized champion of the workers. When searching for the answers to why such significant, new changes could occur in Yankee City, the evidence is clear that economic factors are of prime importance. But before we are content to accept them as the only answers to our problems, let us once more remind ourselves (1) that there had been severe depressions and low wages before and the union had failed to organize the workers, and (2) that the last and most powerful strike which preceded the present one occurred not in a depression but during a boom period when wages were high and economic conditions were excellent. Other factors are necessary and must be found if we are to understand the strike and have a full explanation of why it occurred and took the course that it did.

IV

FROM CLIPPERS TO TEXTILES TO SHOES

1. *The Industrial History of Yankee City*

WHEN we explore the social and industrial history of Yankee City, moving back through the years marked by the beginning of industrial capitalism and through the brilliant years of the Clipper Ship era to the simple folk economy of the earliest community—noticing how an earlier phase of the constantly changing society limits and molds the succeeding ones—it becomes certain that some of the knowledge necessary for explaining the strike can be, and must be, obtained by this scientific process. Furthermore, we see very clearly the times when certain necessary factors which explain the strike appear in the life of the town and how, in conjunction with other causes, their gradual evolution made the strike inevitable. It also becomes abundantly clear that the Yankee City strike was not a unique event but must be treated as representative of a type and that this type is almost certainly worldwide in its importance and significance.[1]

---

1. This chapter was written under the influence of Émile Durkheim, *The Division of Labor in Society* (New York, The Macmillan Company, 1933), and L. T. Hobhouse, G. C. Wheeler, and M. Ginsberg, *The Material Culture and Social Institutions of the Simpler Peoples* (London, Chapman & Hall, 1915). Its orientation is ethnological. The first draft was constructed from original materials gathered within Yankee City, including interviews, only those histories of the town written by local historians, and original documents. When we wrote it we followed a rule laid down at the beginning of the research and stated on page 40 of the first volume of this series: "To be sure that we were not ethnocentrically biased in our judgment, we decided to use no previous summaries of data collected by anyone else (maps, handbooks, histories, etc.) until we had formed our own opinion of the city. In part this was a mistake since it greatly lengthened our field work; in compensation, once we had arrived at our conclusions, we were certain of the facts and operations on which our opinions were founded."

Later we consulted the writings of professional historians and social economists, in particular Samuel Eliot Morrison's excellent works on New England histories and John R. Commons' brilliant study of the shoe industry in the United States. We are deeply indebted to both of these writers, and, in particular, Dr. Commons (see note 2, p. 60). Much of what we say here will be an old story to social anthropologists and economic historians.

The town began in 1635. The colonial settlers of Yankee City, having been tenant farmers in England, founded an agricultural community. For several generations thereafter the colonists continued to perpetuate social and economic patterns essentially similar to those in which they had participated in the motherland; their rural agricultural life centered around the political institution of the town and the religious institution of the church. As in the case of all societies recently transplanted, the community was beset with threats to its existence, both internal and external, and met them by close adherence to traditional modes of life.

The commercial advantages of a location on the river bank were recognized, and the present town site laid out. According to the records, the first wharf in Yankee City was built in 1655. By 1700, interest in commercial enterprises had grown to such an extent that the land along the river, hitherto held in common, was divided into water lots. Maritime commerce was seriously affected by the Revolution, but in the following period, between 1785 and the War of 1812, it revived and the city prospered greatly as a shipping center. The difficulties between France and England following the French Revolution made it possible for American vessels to take over much of the European carrying-trade; Yankee City ships cruised the Atlantic from Baltic ports on the north to the Gold Coast on the south. The economy of the city remained predominantly maritime from the War of 1812 down to the decade of the 1870's. Our local historian records that on one day in this period, a fleet of forty vessels, detained for weeks by easterly winds, set sail from Yankee City. Shipping and shipbuilding reached their apex between the years 1820 and 1865; in the single year of 1854 seventeen new vessels were launched from local shipyards. The rich mythology centering around the exploits of Yankee City ships refers in large part to this glorious period.

Fishing, likewise, reached its greatest importance to Yankee City in the first half of the nineteenth century. Yankee City's historian tells us: "In 1851 there were ninety vessels, measuring 6090 tons, and carrying 985 men, engaged in fishing on the banks of Newfoundland and the coast of Labrador." Within twenty years the number of fishing vessels had declined to thirty and by the turn of the century only a handful of small fishing craft remained. Today the only representatives of that calling are a group of clammers.

The causes of the decline of Yankee City's maritime trade after the middle of the nineteenth century were technological, economic, and geographic. Steam engines were gradually replacing sails, and the increasing size and draft of both sail and steam ships necessitated the use of larger and deeper harbors than that of Yankee City. In addition, improvements in inland transportation made coastwise transportation less important.

Industrial developments as early as the eighteenth century had set the stage for the decline in Yankee City's maritime enterprise. Some of the more farsighted merchants, realizing that the heyday of shipping could not continue indefinitely, began to promote industrial ventures. Yankee City, in its early days, had depended upon England for most of its manufactured goods. During the eighteenth century it began to rely in part upon its own craftsmen, and family enterprises were moved into separate shops. The Yankee City shoe industry can be traced to just such an origin.

As a result of the downward spiral of their maritime commerce after 1850, Yankee City's citizens were again forced landward for their livelihoods, even as their ancestors had been before the days of maritime glory. Cities along the upper river furnished the cue for this change. By harnessing waterpower for industrial purposes, they had already surpassed Yankee City in size. Yankee City had no such natural advantages for development; it had no rapids or falls for waterpower, no tradition of organized industry, and was farther away from the denser population centers than the cities up the river. Although Yankee City set out to adapt itself as best it could to the industrial patterns of nearby communities, it met with only limited success.

From a simple and undifferentiated society, there developed in Yankee City the type of economic life with which standard histories of New England have made us familiar. During the era of shipbuilding, shipping, and fishing, a great number of handicrafts also developed. These included such primary industries as wood-carving, cordage-making, carpentering, blacksmithing, and sail-making. During the winter months the fishermen of Yankee City, as of other New England towns, made shoes. The women manufactured wool and cotton garments within the household. During the nineteenth century, numerous other independent crafts appeared, such as silversmithing, comb-making, leather-tanning, and carriage-building. The most important industry in view of its later development, however, was the manufacture of shoes.

Apprenticeship functioned in the handicraft system of the eighteenth and most of the nineteenth centuries as our trade, engineering, art, and professional schools do in our industrial system today. It was society's educational device for transforming its youth from the "green," unproductive stage to the stage of full economic maturity as master craftsmen. From a history of comb-making, an important craft in Yankee City during the last half of the eighteenth and the first half of the nineteenth centuries, we have the following description of the institution of apprenticeship:

As in most of the handicraft trades the apprentice system was the only entry to comb-making. Its rigorous requirements approached the border line of slavery, for it bound one by legal indenture to serve and obey a master and to be faithful to him in all things. Usually he was allowed two terms of schooling, each three months long. He must not be out after nine, should attend church twice on Sunday and spend Saturday evening preparing for Sunday school. . . . Some of the more generous employers gave their apprentices thirty dollars a year with which to buy clothes. . . . The indenture that bound the apprentice to certain duties also bound the master to certain obligations. He must treat his apprentice kindly, look after his moral character and give him what was known as a freedom suit at his majority.

In the various industries machine methods slowly evolved out of tools and techniques centuries old. Comb-making, in which Yankee City led the country until the middle of the nineteenth century, again offers an example. The first comb establishment in Yankee City and in the country was opened in 1759, using strictly handicraft methods of manufacture. In 1798, the first hand machine to make combs was patented. It was fully forty years later that a comb-making machine run by steam power made its appearance in Yankee City. The first steam-run comb factory was opened in Yankee City in 1842; and in the same year two large "steam mills" for the manufacture of cotton textiles were built. In the next decade there appeared the following establishments: one lace factory, twenty-three comb shops, one woolen yarn factory, one machine shop, one iron foundry, four silver factories, one hosiery factory, and one chair manufactory.

The development of the factory system in Yankee City had three phases, each defined by the dominant industry of the respective periods: comb, textile, and boot and shoe. The comb craft had been founded and developed in Yankee City, and by 1840 there were more than one hundred persons employed in comb manufacturing. During the next five years, seven of the establishments moved to another New England town, which has since become the comb manufacturing center of the country. After 1845 the Yankee City comb industry declined in importance while that of the other town grew. By 1883 only three comb factories remained in Yankee City; by 1893 there were but two; and in 1919, only one—the factory which had been established by the eighteenth-century founder of the craft. Even this succumbed during the depression of 1929–33.

The textile industry was developed during the decade when the comb industry began to decline. This defines the opening of the second phase, which covered roughly the period between 1845 and 1890. The cotton mills never numbered more than five, but in 1853 they employed 1,530 people. The last mill closed in 1930. Two of the remaining mills in this last period moved to the South as part of the general movement of the textile industry out of New England.

As the textile industry lost importance in the economy of Yankee City, shoe manufacturing took its place and employed the major part of the industrial workers. In 1865, there were four shoe shops employing in all about 110 hands. By 1890, about two thousand were employed in the shoe industry, most of them in one factory. This figure of two thousand remained approximately the maximum of employees in the city's shoe industry. During the last seventy years, over forty shoe manufacturing companies have been organized and put into operation in Yankee City, some as reorganizations of units previously operating. In 1945 three shoe factories were in regular operation. These figures indicate both the high mortality rate among the shoe factories and the struggle of the community to maintain its status as an industrial city.

2. From the Cobbler's Bench to Assembly Line—A Shoe History

DURING the first years of the settlement of Yankee City and New England and in the earliest phase of shoemaking, families made their own shoes. The second phase of the first stage was characterized by the itinerant shoemaker who, owning his own tools, made shoes in the kitchen of his customer, using materials supplied by the customer. In this process, the shoemaker was assisted by his customer's family and received his compensation largely in the form of board and lodging. Many families in Yankee City and in the outlying communities, particularly those dwelling on the north bank of the river, became proficient in the art of shoemaking at this stage. They made their own shoes during the winter months, passing down the art in the home from generation to generation. This section of New England has, therefore, a strong tradition of shoemaking.

The next stage began (*circa* 1760) when the shoemaker set up a small shop and made shoes to order for his local customers. These shops were known as "the ten-foot shops," and the customer's order was known as "bespoke." During the first part of this period, the shoemaker still made the complete shoe, but his relation with the market became indirect. The entrepreneur appeared. He was a capitalist shoemaker, hiring workers in their homes to make boots and shoes for him to sell at retail or wholesale. In the second phase of the period the central shop developed where materials were sorted. The parts were cut in the shop, distributed and served in the homes, then collected and the soles joined to the uppers in the shop. Machines were used scarcely at all. The processes of shoemaking were divided, and workmen specialized in one or more operations. Jobs were thus defined within the industry; for the most part, the worker no longer faced his customers.

During this period the market remained local, and the interests of the merchant-master and the journeyman were the same. When improved land and water transportation brought about an expansion of the market, the merchant became an increasingly dominant figure. The bargain became one of price as well as quality, and the interest of the merchant to produce cheaply in order to undersell competitors began to conflict with the maker's desire to earn as much as he could from his labor.

Professor John R. Commons, in an article entitled "American Shoemakers 1648–1895," [2] traced the evolution of the industry in this country from court records of cases involving conflicting interests both within and without the industry. He reports that the first guild of shoemakers, known as the "Shoomakers of Boston," was granted a charter of incorporation in October 1648. Since the days of the "Shoomakers of Boston" other formal organizations have come into existence and left concrete evidence of the various conflicts of interests within the industry.

Before 1852, the menaces to the industry and to the groups within it resulted mainly from the expansion of markets. Until this time all shoes were made by hand, and each craftsman owned his own set of tools. But to meet the increasingly exacting demands of an expanding market, as to both price and quality, it was inevitable that the growing technological knowledge would be utilized to mechanize some phases of shoe manufacture. In 1852 a sewing machine for stitching uppers was invented, and the following decade saw the mechanization of many other processes. This development intensified the split in interests between the owner-control group and the operatives; it also established the subordinate position of the latter which they have occupied ever since. Accelerated mechanization of the industry in the decades after the Civil War occasioned changes in the social structure of the shoe factory.

One of the most important results of the introduction of machinery into shoemaking was the enormous decrease in labor costs. The cost per one hundred pairs was reduced by the machine to well under one tenth of the costs of 1850, and the average labor cost in 1932, we were told, had dropped to forty cents per pair. Another result was the great potential increase in production. For example, an expert hand laster produced fifty pairs a day; a lasting machine, from three hundred to seven hundred per day. A welt machine is fifty-four times as

---

2. John R. Commons, "American Shoemakers, 1648–1895," *Quarterly Journal of Economics*, XXIV, No. 1 (Nov. 1909), 80 81.

fast as welt sewing by awl and needle. The introduction of machines into shoemaking converted it from a strictly hand trade to one of the most specialized of machine industries. The position of labor was greatly modified by the technical revolution. The product has changed only in detail, but the process of manufacture has changed from a single skilled trade, carried on by a craftsman from start to finish, to one of two to three thousand operations in greater part done by machine.

The security of the workers as craftsmen was threatened by the new developments. The shoe workers did not make the machines they were suddenly forced to operate, and they had no way of predicting what jobs would next be mechanized. The owning group had in the machines an effective weapon to lessen the value of the worker's craftsmanship.

Out of this situation arose the Knights of St. Crispin, active from 1868–1872, the most powerful labor organization known up to that time and probably the most important one previous to the modern labor unions. The Knights were organized to protest against the substitution of many "green hands" for the old-time craftsmen, which was made possible by the new use of machines. It was a violent protest, but its life was short.

Since the collapse of the Knights of St. Crispin there have been few effective labor organizations among New England shoe operatives and none in Yankee City until the strike of 1933. Mechanization, however, did not cease, and with it went the subjugation of the workers. Several complete processes of shoemaking were standardized in the course of time. One of them was the "turn" process, particularly adapted to the manufacture of high-quality women's shoes. This process, one of the oldest of modern shoe-building techniques, was standard in Yankee City at the time this study was made in 1931–35.

The turn process has given way, in Yankee City as elsewhere, before the inroads of price competition. Cheaper shoes for women are replacing those made by more complicated and costly processes such as the welt, McKay, and turn methods. Cement and lockstitch processes were evolved to produce shoes that could be sold at a lower retail price. These changes also permit great factory flexibility in adjusting to style variations, an important consideration to the modern manufacturer of women's shoes. The rapidity with which styles change has created rush work demands, necessitating speed in manufacturing processes and a quick and ready adaptability to change. When an order is received, the factory must push production so that the order may be completed before the style changes. With the changing styles, there is a decreasing demand for standardized types of shoes. The result is alternation between rush work and lay-offs. This trend in the manufacture of women's shoes induces a greater than average fluctuation in employment.[3] These factors have contributed to the instability of employment in Yankee City shoe factories.

Another factor in the instability of the shoe factories is the practice of leasing machines. The leasing system was first introduced by Gordon McKay in 1861 and was continued by the larger shoe-machinery companies. The machine manufacturers adopted a royalty system in which the rates per unit of output were the same to both large and small manufacturers. This worked to the disadvantage of the former, who preferred a sliding scale. The small entrepreneur who had been attracted by this feature of shoe manufacture seldom had sufficient capital investment to insure success. The relatively small initial cost of establishing shoe factories resulted in a high mortality among these enterprises.

With the development of a large market in the West and South, the shoe industry has moved many of its production units away from the New England states and closer to the markets. One entire NRA hearing in January 1935 was devoted to a study of the migration of the boot and shoe industry from Massachusetts,[4] and showed that state's share in the national production of shoes to have declined from 47.13 per cent in 1899 to 20.05 per cent in 1934, while its volume of production had diminished—in spite of the increase in national production—from 102 million pairs in 1899 to 71 million in 1934. Some of the important factors contributing to the migration of the shoe industry[5] were the following:

(1) labor disturbances;

---

3. Factories making shoes for men experience smaller seasonal fluctuations and fewer changes in consumer demand.

4. National Recovery Administration, Division of Review, *Report of Survey Committee on the Operation of the Code for the Boot and Shoe Manufacturing Industry* (Washington, D. C., Government Printing Office, July 16, 1935).

5. *Ibid.*, pp. 81 ff.

(2) the necessity to reduce manufacturing expenses and obtain lower labor cost, in order to meet severe price competition;

(3) the location of manufacturing plants in or near the principal markets;

(4) inducements offered to Massachusetts manufacturers by cities and towns located in other states to move to their localities. Such inducements take the form of freedom from taxes, free rent, donations of factory sites and/or property, and, frequently, cash subsidies.

The conditions which we have described (national, state, and local) have placed the Yankee City shoe worker in a precarious position. Changing methods of production and the vicissitudes of the trade itself have led to instability among shoe-manufacturing enterprises. Yankee City is in no position to absorb the output of its factories, and the latter have become more and more dependent on the large chain stores for retail distribution. The number of shoe companies operating in Yankee City and the number of employees have varied from year to year. In 1929, sixteen shoe factories were operating in Yankee City—the largest number operating at one time. The peak in actual employment was reached in 1926 when 2,060 individuals were employed in the shoe factories in the city.

The shoe industry, not only in Yankee City but throughout the country, was one of the first to suffer before the general depression of 1929, showing a decline from a 1923 peak in value of product. During the period of high production, the shoe workers were in a position to dictate their own wages, but during the period of decreasing employment the manufacturers held the dominant position in the internal factory organization and gradually forced down the price of labor.

This pressure, deriving ultimately from retail-price competition, stimulated a concentrated effort on the part of the workers to organize in order that they might resist the manufacturers' desire to reduce costs by reducing wages.

3. *The Strike and the Evolving Social and Economic Systems*

BEFORE we ask ourselves what this economic history has told us about the causes of the strike, let us re-assess our findings. We have spoken of an economic history. However, we do not have one history but several—at least six histories can be traced. We can conveniently divide the technological history of Yankee City's shoe industry into five phases (see Chart I). At least two important stories are to be found here; the tools change from a few basic ones entirely hand-used to machines in an assembly line, and the product from a single pair of shoes to tens of thousands in mass production.

The changes in the form of division of labor (see Chart I) are another story of the utmost importance.[6] In the beginning, the family made its own shoes, or a high-skilled artisan, the cobbler, made shoes for the family. In time, several families divided the high-skilled jobs among themselves, and later one man assigned the skilled jobs to a few men and their families. Ultimately, a central factory developed and the jobs were divided into a large number of systematized low-skilled jobs. The history of ownership and control is correlated with the changes in the division of labor. In early days, tools, skills, and materials were possessed by the family; eventually, the materials were supplied by the owner-manager, and soon he also owned the tools and machines. The sequence of development of producer-consumer relations tells a similar story. The family produced and consumed its shoes all within the circle of its simple unit. Then, the local community was the consumer-producer unit, and ultimately the market became national and even worldwide. Worker relations (see Chart I) changed from those of kinship and family ties to those of occupation where apprenticeship and craftsmanship relations were superseded and the individual unit became dominant in organizing the affairs of the workers. The structure of economic relations changed from the immediate family into a local hierarchy and the locally owned factory into a vast, complex system owned, managed, and dominated by New York City.

With these several histories in mind (and with the help of Chart I), let us ask ourselves what would have happened if the strike had taken place in each of the several periods. In period one, with a family-producing and consuming economy, it is obvious that such a conflict would have been impossible. The social system had not evolved to sufficient complexity; the

---

6. The sequences in the vertical columns of the chart are exactly ordered; the horizontal interrelations are approximations and indicate basic trends.

The Social System of the Modern Factory

Period	Technology	Form of Division of Labor	Form of Ownership and Control	Producer-Consumer Relations	Worker Relations	Structure of Economic Relations
IV The Present (1920–1945)	*Machine Tools* mass production, assembly line methods	Nearly all jobs low skilled; a very large number of routinized jobs	*Outside* ownership and control of the factory (tools leased)	Very few retail outlets; factory merely one source of supply for a chain of shoe stores	Rise of industrial unions, state supervised . . . no (or weak) unions	Center of dominance New York. Very complex financial producer and retail structure. Local factory not important in it
III Late Intermediate Period (approximately to World War I)	*Machine Tools* machines predominate; beginning of mass production through use of the machine (McKay)	A central factory with machines; still high degree of skill in many jobs	First small, and later, large *local* men of wealth own or lease the tools, and machines	National market and local capitalist; many outlets	Craft and apprenticeship (St. Crispin's Union)	Center of dominance local factory; complex hierarchy in local factory system
II Early Intermediate Period (approximately to the Civil War)	*Machine Tools* few machines first application (Elias Howe, etc.)	One man assigns highly skilled jobs to few men; highly skilled craftsmen ("letting-out" system)	Small, locally controlled manufacturers; tools still owned by workers, materials by capitalist, market, controlled by "owner"	Owner and salesmen to the consumer regional market	Informal, apprenticeship and craft relations	Simple economic no longer kinship; worker subordinate to manager
	Hand Tools increasing specialization and accumulation of hand tools	Specialization among several families; a few highly skilled jobs	*Local Control* not all shoemakers need own all tools; beginning of specialization	Local buyer from several producer families sells products (no central factory)	Kinship and neighbors among workers	Semi-economic but also kinship and neighborliness
I The Beginning (early 1600's)	*Hand Tools* few, basic, and simple	All productive skills in the family, including making of shoes; a few cobblers for the local market	*Local Control*—skills, tools, and materials owned and controlled by each family; or by the local cobbler	The family produces and consumes shoes and most other products	Largely kinship and family relations among workers	Very simple non-economic; the immediate family
	Technology	Form of Division of Labor	Form of Ownership and Control	Producer-Consumer Relations	Worker Relations	Structure of Economic Relations

CHART I

The History of the Differentiation of the Yankee City Shoe Industry

forces had not been born which were to oppose each other in civil strife. In the second phase, several families in a neighborhood might have quarreled, but it is only in one's imagination that one could conceive of civil strife among the shoemakers.

In the third phase, however, there appears a new social personality, and an older one begins to take on a new form and assume a new place in the community. The capitalist is born and during the several periods which follow he develops into full maturity. Meanwhile the worker loses control and management of his time and skills and becomes a subordinate in a hierarchy. There are, thus, distinct and opposing forces set up in the shoemaking system. What is good for one is not necessarily good for the other, but the interdependence of the two opposing groups is still very intimate, powerful, and highly necessary. The tools, the skills, and the places of manufacture belong to the worker; but the materials, the place of assembly, and the market are now possessed by the manager. Striking is possible but extremely difficult and unlikely.

In the fourth period, full capitalism has been achieved; the manufacturer is now the owner of the tools, the machines, and the industrial plant; he controls the market. The workers have become sufficiently self-conscious and antagonistic to machines to organize into craft unions. Industrial warfare still might prove difficult to start, although it did occur, because in a small city where most people know each other the owner or manager more often than not knows "his help" and they know him. The close relation between the two often implies greater compatibility and understanding, which cut down the likelihood of conflict. But when strikes do occur the resulting civil strife is likely to be bitter because it is in the confines of the community.

In the last period, the capitalist has become the super-capitalist; the workers have forgotten their pride in their separate jobs, dismissed the small differences among themselves, and united in one industrial union with tens and hundreds of thousands of workers throughout the country combining their strength to assert their interests against management. In such a social setting strikes are inevitable. The remaining chapters will fully demonstrate this proposition about contemporary industrial **America**.

V

THE BREAK IN THE SKILL HIERARCHY

1. *Skill and Status*

WHEN we started our intensive investigation of technological jobs in the shoe factories of Yankee City we expected to find a recognizable hierarchy of jobs based on the degree of skill demanded in their performance. Our assumption was that we would find, in the technological jobs as in the managerial ones, a hierarchy of statuses correlated in some way with the complexity of the task. We expected, too, that the evaluations of technological jobs, both by management, as expressed primarily in rates of pay, and by workers, as expressed in their attitudes toward different jobs, would correlate, in general, with an observable hierarchy of skill.

In the course of our investigation of technological jobs we found that this hypothesis was quite untenable. We discovered that, through the great division of labor and extensive mechanization that has occurred in the shoe industry, there were no high-skilled technological jobs and few medium-skilled; in the modern shoe factory the great majority of jobs can only be classified as low skilled. These findings, as presented in this chapter, also show that comparative wage rates do not even correlate with what little skill differential is left in the modern shoe factory. Some social attitudes toward various jobs do correlate with the relative abilities demanded; in other cases even this correlation has broken down.

We found, too, that the division of labor and mechanization have produced profound changes in the social relations between management and worker by attenuating their relations and emphasizing the conflict of interests between them. The relations between worker and foreman have changed particularly, to the point where the shoe operative now has almost no opportunity to train himself to enter the managerial hierarchy.

The destruction of our original hypothesis, that we would find a skill hierarchy in technological jobs of a modern shoe factory, made it necessary for us to seek further for an understanding of the status hierarchy of the technological jobs within the shoe factory. We found that, instead of the simple and direct correlation we originally expected, there were a number of variable factors which contributed to management's evaluation of jobs as expressed by rates of pay; that these factors also affected the workers' evaluations of jobs but frequently in a manner at variance with the evaluation of management. Instead of a simple, universally agreed upon hierarchy of statuses in technological jobs, therefore, we found the situation to be confused, tense, contributory to suspicion and conflict between workers and management, and disruptive of the social integration of the factories.

2. Forms of Social Control in the Factory

THE controls in the formal business structure of a shoe factory, considered as a management enterprise, are primarily those by which the employer-employee relations are ordered, either by the employer himself or by a subordinate acting as his representative.

An individual agrees with his employer to give his services, mental or manual, in exchange for a prearranged reward in the form of money. An employee, hired either in a supervisory capacity or as a technological worker, must accept the limitations imposed on him by the position of his job in the business hierarchy.

The control hierarchy is clearly recognized in the category of supervisory jobs, for there the individuals are directly affected in the performance of their jobs by the positions they occupy with respect to other individuals. In technological jobs, however, there is no formally recognized hierarchy. The individual worker in each technique exerts his control over objects. The technological workers are at the lowest level in the hierarchy of the business structure. We show later how varying degrees of social prestige may attach to different working techniques on the technological level and may be in part transferred from the job to the individual engaged in it.

Through the functioning of the system of controls, a hierarchical structure of statuses within the shoe factory is apparent. Individuals, by reason of the jobs they hold, are either superordinate or subordinate to other individuals. Besides the control of individuals by other individuals, which we may call supervisory controls, we also observe controls by individuals over objects and materials, which we may call technological controls.

In the production of a large turn-shoe factory employing, let us say, a thousand individuals, there would be a single superordinate head and approximately ten or twelve department foremen. The balance of the employees would be technological

CHART II

Permitted Scope of Activities in Jobs of Different Relative Statuses

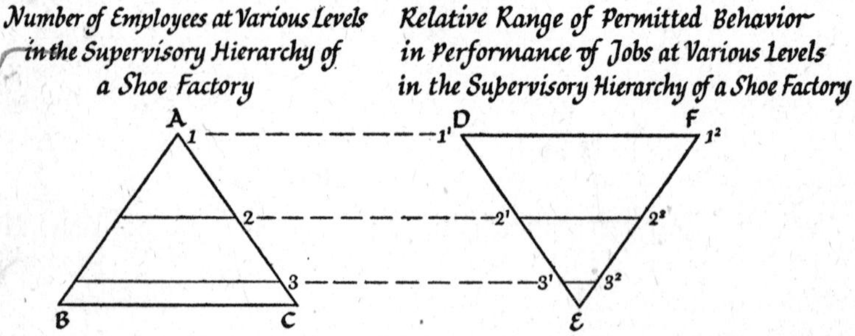

Number of Employees at Various Levels in the Supervisory Hierarchy of a Shoe Factory

Relative Range of Permitted Behavior in Performance of Jobs at Various Levels in the Supervisory Hierarchy of a Shoe Factory

Let us assume that the triangle ABC in Chart II represents the production division of a shoe factory with (1) representing the head of the division, (2) the foremen, and (3) the technological workers.

operatives. Triangle ABC is not drawn to scale, but it indicates that the number of individuals increases with descent in the hierarchy from (1) to (3).

The manager of a shoe factory, by reason of the character of his job, is involved in a more complex set of relations with a greater number of individuals than are any of those subordinate to him; and in ordering these relations he has a greater freedom of choice in his actions than any of his subordinates. The inverted triangle DEF illustrates the variable range of permitted activities in jobs from manager to shoe operative.

The division of labor has become a powerful method of

supervisory control in modern shoe factories. Usually we think of the division of labor merely as the splitting up of jobs into more specific tasks. This leads to certain misconceptions. A job may be regarded as a prescribed set of interrelations between individuals, or between an individual and materials and objects. If a job is divided into two parts, not only are the activities split, but the scope of the interrelations permitted the individuals by the new jobs is curtailed. When an employer hires an individual, he says, "You do this," rather than "You can't do that." The employer's verbal order is in the positive form, but actually he sets up a series of limitations on working behavior. By increasing limitations on working activities, through the division of labor, management of modern shoe factories has made more effective the operation of controls.

Any employer, as a result of his position in the hierarchy of supervisory control, has at hand the means of imposing his decisions upon his subordinates. The right to "hire and fire" and the control of the pay envelope are the strongest of these means. In minor matters, however, wide variations in control exist; and two individuals in equivalent grades of the hierarchy rarely exert the same degree of control upon those beneath them. The differences depend primarily upon the individual attitudes of particular supervisors toward subordinates and the methods they use to control the activities of those beneath them in the hierarchy.

The employer-employee relation is based upon the employment contract. If either party fails to fulfill his obligations, the agreement can be terminated. In theory, therefore, each party exerts a control over the other. But in practice the control by the employer is the more effective, as the employee's need for money is usually greater than the employer's need for the services of any particular individual. This is especially true in the simple mechanical jobs of the modern shoe factory in Yankee City. In this way the division of labor becomes a means of control.

The employment contract, written or verbal, is the strongest evidence of the social control that management exerts. The employer may demand an amount of work and a standard of quality satisfactory to him. The contract further acknowledges the right of the employer to select workers and to decide when any employee shall be dismissed. The exercise of this right is not always unchallenged. In order to equalize the controls in the employment contract the employees often resort to some form of collective bargaining, as we shall describe later. Collective effort has been made to resist the indiscriminate use of the power of management, not only by the labor unions, but also by special groups of subordinates bound in some other mutuality of interests.

The second general type of control observable in a shoe factory is that which the workers in technological processes exert over materials and objects. This control is exercised in different ways, according to the character of the work performed. There are three classes of technology observable in the behavior of shoe operatives: (1) object handling; (2) tool using; and (3) machine operating. In the first class, object handling, the worker controls materials with his hands alone. In each class of technology in a shoe factory the operative functions in part as an object handler. In the second class, tool using, the worker controls materials with the aid of tools which he manipulates. A tool in use becomes an extension of the body of the worker which enlarges the range of his control over materials. There is a vast difference, however, between the techniques involving merely the use of the body and those involving the use of tools. Manipulation of a tool is generally a higher type of technology than object handling because it requires a higher order of muscular coördination. In the third class of technology, machine operating, the worker still remains in part an object handler in his guiding of the material to the machine. The machine operator at times uses tools to regulate and adjust the machine, but he does not use tools to modify the materials directly. He has relinquished to the machine much of his tool-using technique.

The type of machine, as well as the type of tool, varies with the particular process to be performed and the kind of material worked upon. Just as there are hand tools for cutting, holding, and pounding, for example, so also are there machines which perform these operations. The machine eliminates much of the inaccuracy that frequently occurs in hand work and establishes uniformity in the product. The machine operator has little chance to show any craftsmanship in his work. He has slight freedom of choice in his actions, less than tool users and frequently less than mere object handlers.

The highest authority in the local factory in Yankee City was vested in the factory manager. He delegated some of this authority to the works superintendent and the head accountant (in the production and accounting divisions respectively), who were directly responsible to him. The works superintendent, as head of the production division, directly controlled the foremen of the production departments, who in turn controlled the technological workers. The head accountant was in charge of the accounting division and controlled the head clerk. The latter controlled the clerks in the business office.

The manager, as the representative of the owners of the enterprise, directed and ordered employer-employee relations. He was primarily an organizer of the activities of individuals and of factory processes. The ability to organize the complex interrelations of individuals into integrated productive activity was a more important asset in a manager than experience in technical production details. Many managers, however, combined organizing ability with sound technical knowledge.

The activities of a manager varied with the degree to which he delegated authority to his subordinates and also with the degree of authority they were willing and able to assume. Although the line of authority was clearly established in theory, practical usage permitted an individual to present directly to the manager grievances that could not be adjusted by his immediate superior. The manager of a locally owned factory had duties besides the overseeing of actual production activities. He met representatives of suppliers of raw or semi-processed materials and buyers of finished shoes. As a matter of business policy, he had to have suitable facilities for conference purposes. When a factory was merely a production unit of a larger enterprise, however, the manager usually had little direct contact with either suppliers of the major materials or buyers of shoes. Both buying and selling were done elsewhere: the local factory obtained its basic materials through a central purchasing office and shipped its products to a distributing center.

The works superintendent was the head of the production division. His job was primarily to integrate the efforts of the different operating departments through their foremen. His office was generally near the center of factory operations to facilitate conference with foremen. He had to know most of the techniques of shoe manufacture. Although his job was concerned with the organization and control of production operations, frequently the works superintendent, either through a foreman or directly, intervened in the settlement of disagreements among workers. An efficient works superintendent tried to know as many problems of the workers as possible; to know their problems is to know the workers. This knowledge of the workers and their comparative working efficiency aided him in directing the activities of the different departments.

The foremen were in direct control of the technological workers and frequently had the prerogative of hiring and firing them. Foremen were generally selected by the works superintendent with the approval of the manager. Like their superiors in the managerial hierarchy, they were primarily engaged in organizing the individuals under their control. Foremen in the larger departments customarily had one or more assistants, called floor men or floor women.

The foremen were assigned small offices or working headquarters within their departments. If there were departmental stock rooms, the foremen's headquarters were usually there. The foremen were socially closer to the shoe operatives than any other group in the managerial hierarchy. Although a foreman's interest lay with the management, he was more likely than his superiors to have a sympathetic understanding of the interests and needs of the workers. The mechanization of the shoe industry, however, was modifying to a considerable extent the relation between the worker and his foreman; foremen tended more and more to become enforcement officers rather than sympathetic working bosses.

The job of the foreman necessitated constant contact with the workers. His control over the operatives was exerted by different methods, depending upon his personality and the personalities of the workers under him. As a works superintendent put it: "Sometimes a foreman can control workers by a friendly 'pat on the back' or a kind word, and sometimes he must be 'hard-boiled' and ready to use strong language and a strong arm."

3. The Break in the Skill Hierarchy

SKILL is generally related to control over objects through the use of tools, simple or complex. Its technological import is recognized by application of the term to persons showing technical ability as distinguished from those whose prestige is derived solely from mental ability. For example, we may speak of a *skilled* surgeon or dentist, but we are more apt to refer to an *able* physician or lawyer. This limited application of the term, skill, is confusing since any technique can demand skill in its performance if an operative has some freedom of choice in the manner in which he performs his task.

Skill is an attribute of a person, but by convention the word is also applied to jobs. The use of the term as an attribute of a job may be confusing. Obviously, a job of itself cannot be skilled. The idea of skill involves its accomplishment; the accomplishment involves an individual. It is the individual who possesses the skill. Strictly speaking, a job cannot even be said to demand skill. The job can be done in some fashion, or at least begun, by an unskillful person. The only definite meaning the term can have in an expression such as "a skilled job" is that the job affords an opportunity for the exercise of skill in its performance. A job is skilled if the person who undertakes it can put his skill to use in its accomplishment.

Although skill is a valuable concept, it is not an easy one to use in comparisons; for it is generally associated with a sequence of actions, and different sequences of actions are not easy to compare unless they happen to be of the same type. An observant foreman is able to make a very fair estimate of the expertness that can be shown in any one type of job under him. Only with difficulty, however, can he make comparisons of the expertness that can be shown in different jobs in his own department.[1] If he compares the expertness that can be displayed in the performance of jobs in his department with that permitted by jobs in other departments, he is usually unable to justify his words by sound reasons.

---

[1] In interviews, a stitching-room foreman, a factory executive, and a union official selected from the numerous jobs in the stitching departments the same four jobs as the most skilled; but each person interviewed advocated a different ranking among the four.

Technological skill can be considered a measure of the control over objects that a job affords an individual. But before endeavoring to locate a hierarchy of jobs, such as those of shoe workers, on the basis of the skill which their efficient performance would require, we must discover some of the common factors in the expertness allowed by jobs involving different techniques. Then, by using these factors as measuring units, a uniform basis of comparison may be set up which is applicable to all such jobs in the factory.

Within limitations, a skillful person chooses what to do, decides how to do it, and then does it dexterously. Selecting and deciding plus dexterity in accomplishment are characteristics of skill. We shall apply the term "high skill" to jobs permitting the exercise of judgment and the making of decisions as well as dexterous accomplishment. A job is high skilled to the extent that it permits freedom of choice to the worker and necessitates, for its efficient performance, reasoned selection from different modes of action. Further, we shall designate as "low skilled" those jobs of a routine nature in which the worker functions according to a set pattern that is definitely prescribed but may be executed with dexterity. Those jobs which are to a large extent prescribed but which still permit some freedom of choice to the worker we will term "medium skilled." No type of skill can be measured satisfactorily in terms of another type. Chart III demonstrates the different character of the three types of skills in technological jobs and shows that proficiency in the different types cannot be compared.

In the days when shoemaking was a handicraft there was a highly developed hierarchy of jobs based on skill. The hierarchy, moreover, was functionally arranged so that a raw apprentice entering upon his training period performed low-skilled jobs and worked at them until he gained some degree of proficiency (see Chart III). In these jobs he began to train his hand and eye so that when he had mastered the low-skilled jobs he was prepared to begin learning the medium-skilled ones. Gaining proficiency in these journeyman jobs, in turn, prepared him to undertake the high-skilled jobs of the master craftsman. When he had attained proficiency in the high-skilled jobs he was a master shoemaker and stood at the top of a technological hierarchy where great prestige was at

CHART III

Distinction between Types of Skill in Jobs and the Degree or Extent to which Proficiency May Be Developed by Operators within the Varied Classifications

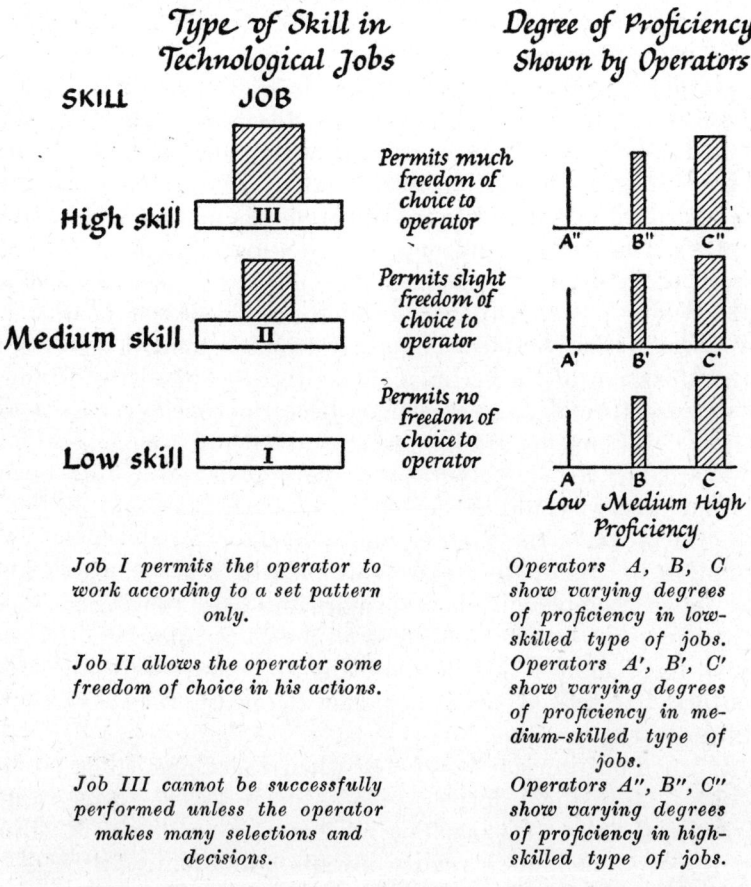

Job I permits the operator to work according to a set pattern only.

Job II allows the operator some freedom of choice in his actions.

Job III cannot be successfully performed unless the operator makes many selections and decisions.

Operators A, B, C show varying degrees of proficiency in low-skilled type of jobs.

Operators A', B', C' show varying degrees of proficiency in medium-skilled type of jobs.

Operators A", B", C" show varying degrees of proficiency in high-skilled type of jobs.

Great proficiency in a low-skilled job is not equivalent to slight proficiency in a higher type because the types differ in character.

tached and where he was economically secure because the services of master shoemakers were always in demand and the supply was limited because of the long period of training required to attain the position of a proficient master craftsman. Also, when a man achieved high status in the technological hierarchy he acquired supervisory status as well, for master craftsmen were also working bosses, overseeing the work of those lower in the hierarchy.

The situation of the individual technological worker in a modern turn-shoe factory is quite different from all this, we discovered. No longer is it possible for him to start in low-skilled jobs and progressively prepare himself for higher-skilled jobs. The loss of this opportunity to the worker is mainly attributable to two related trends in modern shoe manufacture. One is the tendency toward greater and greater division of labor, which means that individual jobs are broken into two or more components which are thereafter performed by different individuals. The other trend is toward increasing mechanization of the technological processes which tends more and more to make the workman perform routine operations. These developments are interrelated and reinforce one another.

Watching the shoe operatives working in the techniques of shoemaking, we could not fail to recognize that as new machines were installed in the factory more and more of the tool-using function of the operatives was absorbed by the machine and the job of the operative became more and more routinized. Real craftsmanship lost its usefulness as the operatives who had had much freedom of action in tool-using techniques were forced to conform to a set pattern of behavior attuned to the rhythm and tempo of the machine.

It stood out in clear relief that when a machine is built it becomes a mechanical device which of itself has no social value. The social value of a machine is derived from its use. The inventor of a machine in reality invents a mechanical process in which a machine and an operator interact in a prescribed way. The inventor of the process defines the working behavior of the operative in his relation to the machine. To a large extent, the operative is denied freedom of choice in his actions. The denial of opportunity for the use of any creative initiative by workers, bound so closely to their machines in their daily tasks, has had a marked effect upon the operatives.

The relation between an operative and the machine at which he works may be called a mechanical relation—in essence it is not a social relation at all. But when the operative's working behavior is almost completely controlled by the functioning of the machine he serves his attitudes toward it are likely to change. Instead of regarding the machine as merely a mechanical object, part of the factory equipment, the worker tends to assume a proprietary attitude toward it, to refer to it as "my machine," etc.[2] Not only does the worker develop proprietary attitudes toward the machine, but he tends to personify it and to develop quasi-social attitudes toward it. These various sentiments on the part of the worker tend to integrate him closer and closer to the machine at which he works.

Anything that disrupts his integration with his machine or the technique of operating it is likely to be most unwelcome to the worker. Even changes intended to increase his comfort or safety often arouse his indignation. This is probably because, since the worker learns how to operate his machine merely by practicing, and any change in the simple routinized technique requires re-learning on his part, such a change seems to the worker to threaten his security and may, in fact, actually do so. At times, a considerable part of the friction between operatives and management in a modern shoe factory springs from this source. Workers' reactions to innovations are frequently surcharged with emotions that are incomprehensible to management.

The attitude of the worker that he is socially integrated with his machine probably tends to change the character of his social relations with other workers during working hours. We further observed that operatives working on machines had far less opportunity to converse with their co-workers than opera-

2. A foreman described an incident illustrating this point. A girl who operated a machine had been sick for a time and had not been able to work. The foreman said: "Of course I put another girl on the machine she used. She came back and said, 'Where is my machine?' I said, 'Your machine? Did you buy it? Maybe you would like to. Maybe you could buy the factory, too. What do you mean, your machine? You were out sick, weren't you? You take the machine I give you.'"

tives working alongside each other at benches or tables. Even though the spatial distance between operatives may be small, as in the case of the operatives of sewing machines in the stitching department, the exactitude of the demands of the machine precludes conversation while working. As soon as the bell rings for noon hour or at the end of the working day the stitching room reverberates with voices of machine operatives.

The operatives of the larger machines in other parts of the factory are sufficiently apart from their co-workers to make conversation impossible. All this is in marked contrast to the continual hum of conversation noticeable among all bench or table workers engaged in hand operations.

The work of the latter may be in many cases simple and monotonous, but the freedom to converse may relieve much of the strain from the work and make it less exacting. Thus, the noise many machines make, and the spatial isolation of workers that is a frequent result of mechanization, both contribute to the social isolation of workers from one another during working hours. The effect of this is to prevent social solidarities from developing between workers or to lessen whatever solidarity has developed prior to mechanization. Mechanization is, for this reason and others we shall mention shortly, valuable to management as a means of control over workers. While machine processes were adopted by shoe factories primarily to reduce costs and to speed the processing, the machine has other great advantages over the human worker from the managerial point of view: its performance (barring breakdowns) can be predicted with certainty, and a machine presents no problems of a disciplinary nature. The forces which govern human behavior are little known, and the maintenance of productive activity is always more or less problematical when management has to deal with socially integrated groups of workers. Control problems are simplified, therefore, on two counts through mechanization: (1) machines are easier to control than human beings, and (2) mechanization tends to disrupt the social solidarity of the workers, who thereby became easier to control than they would be if they were able to maintain close social relations during working hours.

One of the specific points at which worker-management relations are most directly affected by mechanization is in the relations between workers and their immediate superiors, the departmental foremen. With the establishment of the set pattern of working behavior which results from mechanization, the foreman does not need to be a working boss interested in improving the technical ability of the shoe worker. In the handicraft days journeymen or master shoemakers directed the activities of apprentices and corrected their errors, instructing them in the craft of shoemaking. The foreman of a mechanized department in today's shoe factory need not have had long training in shoemaking techniques in order to supervise simple mechanized operations. He can be selected by management purely for his ability to enforce a prescribed, set pattern of working behavior. Sometimes men are even taken from other industries to be foremen in the shoe factories. This new relation between workers and their supervisors accentuates the division of interest between the two groups. Workers today tend to feel, with considerable justification, that even their immediate superiors may not understand or care about them. The social situation is quite different from the one that existed when the technological worker's boss was a man who had always followed the same course as his subordinate until, after years of training, he had become a master shoemaker.

Moreover, in the handicraft days every man learning the shoe trade could look forward with justifiable hope to the day when he might be a working boss himself. With the advent of mechanization, however, the changed nature of the relations between worker and foreman has largely destroyed the worker's chances of getting into the supervisory hierarchy. As a machine operator he has little chance to train himself in the techniques necessary to a modern foreman. Even if he were capable of doing a foreman's job, he has no opportunity to demonstrate the fact. The modern operative is virtually condemned to seek his security and his working prestige strictly in the working techniques.

The machine operator's chances of enhancing his prestige or his security through his technological ability are very small in the modern shoe factory. Nearly all the machine jobs entail the set pattern of working behavior which characterizes the low-skilled job. Some individual operatives become surprisingly expert at their low-skilled jobs, spurred on to speed by the piece-work basis of pay. But greater proficiency in one mechanized, low-skilled job does not prepare the modern worker to do any other mechanized job. Even if it did, the fact would be of little value to the operative because the job for which he might prepare himself would be another of low skill, offering no greater security, pay, or prestige than before. There is, in other words, no skill hierarchy in the mechanized processes of the modern shoe factory through which an operative may progress as his abilities develop. In addition, there is the constant threat to the operative's security that the machine process at which he has attained proficiency may be discontinued in favor of some other process; he has no control over such technological changes. Again, even at best, nothing but a few days' or a few weeks' practice protects a proficient operator in one of the simpler machine jobs from the large group of unqualified workers who seek employment. All of these factors tend to increase the subordination of the individual worker to management; from the management viewpoint they are valuable means of social control over workers.

Low-skilled jobs in the factory are not limited to machine jobs. Hand operations, too, through the great division of labor which has taken place in modern shoe manufacturing, have in many cases been split into simple, standardized operations. So far has this division gone that no technological jobs remain, either hand or machine, which by our classification could be rated as higher than medium-skilled; the great bulk of jobs belong definitely in the low-skill category. The ability an individual acquires at one low-skilled job, whether it be a machine or a hand job, ordinarily does not fit him to do another job. Even if it did, no advantage would derive to the worker because in hand and machine jobs there is no longer a technological hierarchy through which he can progress to a position of greater security and prestige. Chart IV graphically shows the leveling of jobs which has occurred through the division of labor and mechanization.

4. *Social Effects of the Break*

WHAT, then, has become of technological skill in shoemaking? With the constant demand for better shoes at lower prices there would seem to be need for greater creative effort and manufac-

The Break in the Skill Hierarchy 81

turing flexibility than ever before. This would, in turn, suggest that there should be more high-skilled jobs in modern shoe manufacture than ever before. The answer is that these high-skilled jobs do exist—they may even be of a higher order of skill than master-craftsman jobs in the handicraft days—but

CHART IV

The Result of the Leveling of Technological Jobs in the Shoe Factory

Hierarchical arrangement of jobs in the days of handicraft shoemaking. The individual's security and prestige increased as he progressed upward from job A to job D.

The common level of nearly all technological jobs today, showing the breaking up of each old job into several simple ones (division of labor). Modern operatives are nearly all at the same low level of prestige and security because there is little difference in either of these respects between any of the jobs from A_1 to D_4.

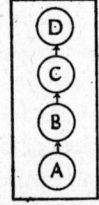

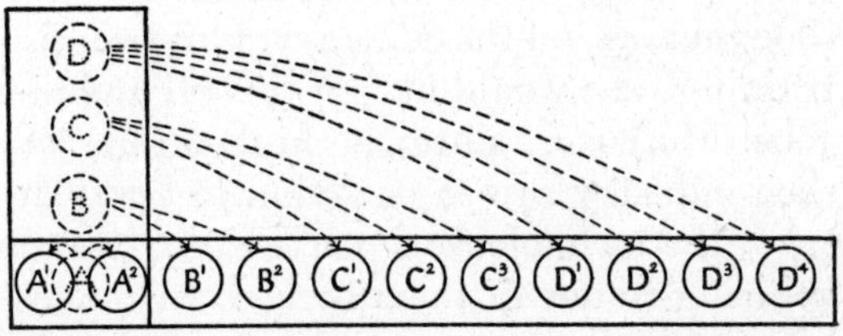

The upward pointing arrows imply the preparation an individual got by working in one job for doing the next higher job in the hierarchy.

The lack of arrows connecting the jobs A_1 to D_4 implies the fact we discuss in the text, that working in one job does not prepare the modern operative to do any other job.

they are *not* to be found in the shoe factories today. They exist in allied industries—the manufacture of shoe machinery, for example. Designers and engineers in this field invent new and cheaper ways to make shoes and design machines to perform the new processes. Since the shoe-factory workers holding high-skilled jobs are a potential threat to management's control of shoe operatives, inventors apparently are encouraged to break down complex jobs into series of simple, easily standardized operations. An important result of their work, therefore, is to eliminate more and more of the skilled jobs from shoemaking, tending to accelerate the leveling of technological jobs in the shoe factory to a common low order of skill. To a lesser extent, research departments in other industries (chemistry is a case in point) also reduce the number of high-skilled jobs in the shoe factories by developing new substances which simplify shoemaking. Designing departments and the skilled jobs in connection with them have been almost eliminated from modern shoe factories. Pattern makers, whose training is obtained outside of the shoe factories, take the place of designers in the factory.

This change in training opportunities offers a clue, of special significance, to the social organization of modern shoe factories. All high-skilled jobs require special training of the individuals who hold them: the designers of shoe-manufacturing machinery, for example, are trained in engineering schools; research chemists have special university training; and pattern makers are trained in schools of design. The operative in the shoe factory, on the contrary, has no opportunity to acquire the training that would fit him to hold any one of these high-skilled jobs. Moreover, since the high-skilled jobs connected with the shoe industry are to be found in separate though allied industries, the technological worker in a shoe factory is barred from aspiring to one of them by the interpolation of two managerial hierarchies between his low- or medium-skilled job and the high-skilled job in another company. The management of the shoe-machinery companies directs the making of machines, and the top management of the shoe-manufacturing companies orders the installation of the mechanical process in the shoe factory.

Even the workers in high-skilled jobs, the research men in industries allied to shoemaking, today occupy a position of much greater subordination than that formerly occupied by the master craftsmen in the shoe industry. For, although they have absorbed nearly all of the high technological skill involved in the production of shoes, they have no control over the use of the machines they assist in creating. The machines, new chemical substances, designs for shoes, and all the rest are useless to the inventors until they are sold (or rented) to concerns that manufacture shoes. This transfer from one company to another is carried on between the managerial hierarchies of the two companies; neither the holders of high-skilled jobs in the allied industries nor the holders of low- and medium-skilled jobs in the shoe factories have any control over the agreements made between them.

The gradual elimination of craftsmanship from shoe manufacturing has had far-reaching effects on the type of personnel attracted to technological jobs in the shoe factories, on the type of individuals whom management wants for such jobs, and on the relations between management and worker. The limitations that have been placed on working behavior restrict the worker in making any satisfying use of his individual mental or manual abilities. Workers ambitious for individual betterment are frequently frustrated in the shoe factories of today. This often makes them discontented and critical of superiors. Educated persons particularly are prone to such discontent. Management is aware of this and attempts to overcome the difficulty by discriminating against educated persons who seek technological jobs. Some foremen (who frequently do their own hiring and firing) are quite conscious of their own discrimination. One of them said: "We don't want educated workers. Educated workers are no good." Management wants workers who will do simple unskilled jobs without questioning the authority of superiors.

Nevertheless, an effort is still made by management and workers to rank technological jobs in a status hierarchy. Management feels a need for such ranking in order to justify wage differentials in various technological jobs; and operatives cling to the ideology that theirs is a skilled craft. This traditional view of the latter is flattering to self-respect, giving the workers a sense of pseudo-security and prestige. It is an attitude they find hard to relinquish even though reason shows it to be obsolete. The workers' view of shoemaking as a skilled craft does not require them to regard all factory jobs as skilled. Certain jobs only, the modern equivalents of formerly high-skilled jobs, still wear this aura of high prestige among workers, though often the skill differentials have actually diminished to the vanishing point.

We shall conclude this section by discussing the present attitudes of workers and management regarding the relative skill of various jobs in the factory and comparing them with our own disinterested classification of the skill characteristics of the various jobs.

We found a positive correlation between some of the lowest-skilled jobs and the low pay and low prestige of the jobs.[3] All the jobs in the assembling department and most of those in the packing department, for instance, consisted of the simplest sorts of object handling. Most of such jobs are on a time basis of pay. Hourly earnings for men in both departments (excluding the treers in the finishing department) were about twenty cents per hour below the factory average (for men) of 59.5 cents per hour. Jobs in some other departments, such as leather heeling and most of the finishing, were about average for the factory in complexity and skill and also approximated the factory average in pay.

Some jobs, if not actually classifiable as medium-skilled jobs, at least required care on the part of the operative. On them may depend the ultimate fit and wearing qualities of the shoes. One such job was that of the channelers in the sole-leather department who cut the channels for the Goodyear stitching operation. The pay of the channelers averaged thirty-nine cents an hour above the factory average for men. Goodyear stitching was another important job since a poorly stitched shoe will neither fit well nor give good wear. Goodyear stitchers were the highest paid of all factory workers, earning nearly twice the average pay for men. Edge-setting and edge-trimming of soles in the finishing department were jobs which required careful work because the rotating knives and oscillating hot irons used in these operations on nearly finished shoes could easily mar or ruin them. Workers holding these jobs averaged a little over 67 cents an hour, about eight cents above the factory average for men. These jobs were regarded by workers as the most skilled, and the better-than-average rates of pay reflected, in part, management's evaluation of the importance of the jobs.

It is a striking fact that some of the least skilled jobs in the factory, by our classification as well as by the consensus of

---

3. The earnings of shoe operatives as reported here and later in this book were derived from figures submitted to us by factories in the first half of 1936. The processes in different departments of the local factory will be briefly described in Appendix D

opinion of workers and management, were paid well above the factory average. This was true of jobs in the sole-leather department other than those of the channelers. Although these jobs were repetitive and monotonous, they afforded the workers an opportunity for tremendous speed-up. The sole-leather department, exclusive of channelers, averaged nine cents per hour more than the factory average for men. Other examples occurred in the stitching department where some of the girls working at extremely simple jobs, such as stamping, earned more than anyone else in the department. Again these were jobs which were capable of great speed-up. Earnings of many of the stitchers, on the other hand, especially fancy stitchers, were low even for women operatives, who averaged in general twenty cents an hour below the average for men. The explanation given for the low pay of stitchers was that styles change so rapidly that the girls could not work up much speed at any one style. Another factor entered here: the speed of stitchers was limited by the tempo of their machines while the simpler, higher-paid jobs, like stamping, were often hand operations without mechanical limit on speed-up. Then, too, the stitching operations were so varied that management admitted great difficulty in determining proper basic piece-rates. As a result, there tended to be in this department an inverse relation between complexity of jobs and rates of pay. Some of the lowest-skilled jobs paid the most, and some of the most complex jobs paid the least.

According to community and worker traditions, cutters are the aristocrats of the shoe trade. Actually, cutting is not a particularly skilled job in the factory today because skins are well sorted before they are delivered to the cutting room; therefore, the operatives no longer need the thorough knowledge of leather they once had. Nevertheless, cutting is definitely a higher-skilled job than most sole-leather operations and compares favorably with such highly paid jobs as channeling, Goodyear stitching, and edge-setting and trimming. Despite this fact, cutters averaged in earnings over a cent an hour below the male average. Bases of evaluation other than skill entered into the rate of pay for cutters.

The two remaining jobs in a turn-shoe factory which still require some craftsmanship on the part of operatives are making and wood-heeling. Makers, particularly, must be proficient with a greater variety of hand tools and techniques than any other workers in a turn-shoe factory. Moreover, these blacksmiths of the trade perform exceedingly arduous work, and the making operation must be done more carefully now than ever before. Makers formerly could turn out eighty pairs of shoes a day; but, to match the quality of machine-lasted shoes, their output has been reduced to only about thirty pairs. In spite of all these considerations, the makers earned only 57.8 cents per hour, 1.7 cents less than the factory average for men. The principal explanation for this lay in the disparity between the supply of qualified makers and the demand for their services.

Wood-heeling is a somewhat less skilled job than making; but, except for the makers, wood-heelers had to master more hand techniques and use a greater variety of tools than any workers in the factory. Wood-heelers' earnings averaged 12.8 cents per hour over the factory average for men and thus correlated with the skill characteristics of their jobs much better than did makers' earnings. The special explanation for this fact lay in the solidarity of the workers in this department.

As we have shown, there has been a break in the skill hierarchy in modern shoemaking; shoe factory operatives are limited to low-, or, at most, medium-skilled jobs in the technological end of shoe manufacturing. This break has come about through the great division of labor and extensive mechanization that have occurred in the shoe industry and through the removal of all high-skilled jobs connected with shoemaking to the research departments of allied industries. These same two factors, the division of labor and the effects of mechanization, have brought about a fundamental change in the social relations between workers and management. Instead of a working boss (foreman), trained in a complex technology of shoemaking, and a managerial hierarchy with a similar background, the workers' supervisors are becoming mere enforcement officers and disciplinarians, forcing set patterns of working behavior on the operatives. This new relationship accentuates the split in interests between workers and management. It also tends strongly to prevent workers from leaving the technological end of shoe manufacturing to enter the supervisory field. Finally, there is confusion in the value systems of the workers and management for there is no consistent correlation between skill of job and rate of pay in the technological jobs of the modern shoe factory.

5. The Strike and the Break in the Skill Hierarchy

WE believe that the break in the skill hierarchy contributed importantly to the outbreak of the strike, to the course it took, and, in particular, to the coming of the union. The hierarchy of crafts which once organized the relations of the workers and provided a way of life for the shoe workers was really an age-grade system. Youngsters served their hard apprenticeship and, as neophytes, learned their task; even more importantly they were taught to respect the skills they had learned and those they looked forward to learning. Above all, they acquired respect and admiration for the older men above them who had acquired the skills and who occupied the proud positions of journeymen and master craftsmen. These youngsters aspired to achieve for themselves a similar high position and respect. Each young man, in direct face-to-face interaction with those above him, imitated and learned a way of life while being highly motivated by the strong desire to escape the irksome limitations of his present low position and to attain the higher place where he would have the satisfaction of making his own decisions and possess the prestige and pay consequent to such great eminence. By the time he had learned how to do the things needed to equip himself for advancement, enough time had passed to mature him sufficiently to act the part of a man. There can be little doubt that age factors as well as those of skill determined the time for advancement.

During this preliminary period he learned that he was a craftsman, with a particular place in the whole system, and that there were responsibilities and obligations he had to learn which would give him certain rights and privileges. Thus, while he internalized this behavior and all its values and their many subtleties and learned what he was as a man, he became an inextricable member of the honorable fraternity of those who made, and who knew how to make, shoes. In this system, workers and managers were indissolubly interwoven into a common enterprise, with a common set of values. In this system the internal personal structure of workers and managers was made up of very much the same apparatus, and their personalities were reinforced by the social system of shoemaking.

In learning to respect the skill of the master craftsman, the apprentice learned to respect himself. He had security in his job, but he had even greater personal security because he had learned how to respect his job. And because he was a member of an age-graded male fraternity made up of other men like himself who had the knowledge and necessary skills to make shoes, he possessed that feeling of absolute freedom and independence and of being autonomous that comes from being in a discipline. He spent his life acquiring virtue, prestige, and respect, learning as he aged and climbed upward, and at the same time teaching those who were younger than he and who aspired to be like him.

Slowly this way of life degenerated and the machine took the virtue and respect from the worker, at the same time breaking the skill hierarchy which dominated his occupation. There was no longer a period for young men to learn to respect those in the age grade above them and in so doing to become self-respecting workers. The "ladder to the stars" was gone and with it much of the structure of the "American Dream."

When the age-grade structure which organized the male aborigines of Melanesia and North America into a hierarchy of prestige and achievement was broken under the impact of white civilization in many of these societies, the frustrations suffered by those who had once known respect for themselves and others crystallized into aggressive movements or into attempts to abolish the new ways and to retreat into the old and cherished ways of the past. There are many resemblances between what happened to these simple, non-European societies and what happened to the craft hierarchy of Yankee City.

The parallel between Yankee City's age-grade structure and theirs cannot be pushed too far, but certainly the two share obvious characteristics. In the earlier days of the machine, the Knights of St. Crispin was organized and attempted to stop the further introduction of machinery (see Chart I). Most of the members longed for the good old days when there were no machines—when a trained hand and eye did the whole job. These attempts failed and the organization collapsed because it was not adaptive and could not stop the inevitable advance of our industrial technology.

When the whole age-grade structure of craftsmanship had almost entirely collapsed and the American shoe worker was thereby denied his share of the "American Dream" he and his kind were ready for any mass movement which would strike at those they charged, in their own minds, with the responsibility for their present unhappy condition. Much of this behavior was not conscious. Much of it was feeling rather than thought as indeed it had been in the mass movements of the aboriginal Melanesians and North American Indians. It seems certain, however, that American workers, taught from childhood that those who apply themselves to their craft and practice the ethics of the middle class would be rewarded by achievement and success, would rebel and strike back out of sheer frustration when they found out that the "American Dream" no longer was attainable for them and that the hard facts belied the beautiful words they had been taught. It seems even more likely that the effects of the break in the skill hierarchy were potent forces which contributed their full share to the workers' striking and the union's becoming their champion.

VI

WAGES AND WORKER SOLIDARITY

1. Women, Wages, and Solidarity

HAVING discovered that the degree of skill of a job bears no predictable relations to the rate of pay or to the evaluation of the job by workers or the community, we found it necessary to seek the various bases on which management, the workers, and the community made their evaluations of jobs and workers. These must be known and demonstrated before the internal social structure of the shoe factories can be fully understood. It is also necessary to understand the external structure of the factories, for this structure is today dominated to a large extent by conflicting values.

A fact immediately noticed when pay rates are examined is the discrepancy between the earnings of men and women.[1] The mean rate for all male operatives in the factory was 59.5 cents per hour; for all women, 40.5 cents per hour, only a trifle over two-thirds of the men's average. This same differential was maintained in the stitching department where the 306 women operatives averaged 41.1 cents per hour, and the seven men, 62.5 cents per hour.[2] In the packing department, on the other hand (eliminating the treers who were men on piece work), men and women were paid on a time-basis average of 39.6 cents per hour; both men and women were paid at the "women's rate" in this department.

Women's pay was less than that of men because there was a sexual division of labor in the shoe factories. Women were employed in jobs which required little physical strength. Their jobs were primarily evolutions of tasks which they have traditionally performed in the home—fitting, stitching, washing,

---

1. The classification of workers by sex, age, ethnicity, social status, etc. is based on the analysis of 985 shoe operatives as of 1935. Further correlations of these characteristics will be made later.

2. This discrepancy is probably largely accounted for by the greater speed of the men, as the workers were paid on a piece-work basis. There was no basic differential in piece-work pay between men and women in similar jobs.

cleaning, and packing. Particularly, women were employed in simple machine jobs which did not require great physical exertion. In our typical Yankee City shoe factory there were twice as many women machine operators as men, even though 57.5 per cent of the factory employees were men. As the division of labor and mechanization increase and complex hand jobs are broken up into simple machine operations, women are coming to be employed in a wider variety of techniques in modern shoe factories than ever before. This sexual dichotomy is recognized by workers as well as management and a certain stigma attaches to a job that is recognized as "women's work." Thus management can pay low rates to women employees without arousing any great resentment on the part of workers, male or female.

This consideration is a strong incentive to management to hire as many women operatives as possible, since production costs can be lowered considerably by reducing wages. Another characteristic of women operatives also predisposes management to hire them in preference to men where possible: the greater ease with which women can be controlled. Although foremen seem to feel that women as individuals are harder to supervise than men, because they are more given to petty annoyances and indispositions, still they feel that large groups of women operatives do not constitute the potentially serious threat to the operation of controls that men operatives do. This is because women seem not to develop the social solidarities in their working relations that men do. It was with reference to the qualities of low labor cost and ease of control that one of the Yankee City shoe factory executives said: "Women in general are more desirable than men as stitchers, but for those particular operations which require special care men are preferable."

Though a minor point, this is one on which not all operatives were agreed. A woman stitcher of long experience said: "Men are faster than women and are better in certain jobs which require the hardest manual labor. Women are more conscientious and more careful of details." This reflects a woman's attitude toward the work done by members of her own sex. There is no doubt, however, that men operatives in general, and probably many of the women operatives too, regard women's jobs in the factory as somehow intrinsically inferior to those of the men. It is certain that the few men stitchers in the stitching department were looked down on as holding "women's jobs."

Women were thought by the average worker to be less capable than men. For instance, an interviewer asked a cutter why there were no women in the cutting room. He replied: "A woman couldn't do that because she couldn't handle a knife. She'd start cutting out things and before you know it she'd cut her fingers." Many of the allegedly inferior jobs held by women, as we have said, were simple mechanical ones resulting from the breakup of former more complex hand jobs. Yet equally simple machine jobs held by men did not carry the same stigma of inferiority in the eyes of the operatives. On the other hand, foremen reported that men often complained that women workers always have the easy jobs, always sitting, while the men have to stand up and work hard.

Lack of social solidarity among women operatives was mentioned above as one of the managerial considerations that increases their desirability as employees. From the point of view of the technological workers, however, social solidarity among themselves is very important. Workers can present demands much more effectively—whether for wages or other improvements in working conditions—if the demands are made by a cohesive group rather than by individuals with separate representations. We found some illuminating correlations between the degree of social solidarity of working groups and such matters as rate of pay, self-government of the group, and the like.

2. Ethnic Groups and Solidarity

PERHAPS the outstanding case of a high degree of worker solidarity was found in the wood-heeling department. Of the forty-five men in this department, thirty-three were native Yankees and twenty-six of these were "Riverbrookers." The Riverbrookers were the most "clannish" and socially solid group of workers in the shoe factories. Most of them resided in the separate community of Riverbrook where they had isolated themselves from, and been isolated by, the other Yankees. Even Riverbrookers who had moved into Yankee City maintained their ties with their native community and took little part in the general social life of Yankee City. Their social solidarity was greater than that of any ethnic group and far surpassed that of the Yankee City natives. Being in the majority in the wood-heeling department, this extremely well integrated group controlled the department to their own advantage in many ways. For example, having strong prejudices against members of the newer ethnic groups, they were able to prevent members of these groups from getting jobs in the department. A factory manager said that if a Greek, for instance, were to be given a job in wood-heeling, the Riverbrookers would make life so unpleasant for him that he would be forced to quit. In the matter of wages, too, the solidarity of the Riverbrookers appears to have stood them in good stead, since the earnings in this department averaged 12.8 cents per hour over the factory average for men (see Table 1) and well above the earnings of any other large department. Management had made no attempt to adjust rates downward in the wood-heeling department; apparently the saving that could be effected by doing so, even where so many men were concerned, was not sufficient in the eyes of management to warrant risking the disruption of production that would occur if the wood-heelers had taken concerted action against management. Thus the social solidarity of the workers in this department gave them considerable bargaining power in advancing their interests over those of management.

The edge-setters and edge-trimmers possessed a solidarity somewhat approaching that of the wood-heelers. The proportion of Riverbrookers was not so high, but half of these workers were native Yankees and the other half French and Irish. There were no representatives of the newer ethnic groups in this work. The foreman, a Yankee, stated flatly: "We don't want foreigners in this department." The social solidarity of this group of workers apparently had an effect on wages, too, for the eighteen edge-setters and trimmers averaged 7.6 cents per hour over the factory average for men.

In sharp contrast to the wood-heelers and the edge-setters and trimmers, both in social solidarity and in average earnings, were the operatives in the leather-heeling and making departments. In leather-heeling over half the workers were ethnics, representing several of the more recent ethnic groups. This department was notably lacking in ethnic solidarity and did not display the antagonism to "foreigners" so apparent among the two groups previously considered. The earnings of the eighteen leather-heelers approximated the factory average for men.

The making department, largest completely male depart-

TABLE 1

Classification of Jobs, and Groups of Jobs, Showing the Number of Male Workers Employed in Them and the Difference between Their Average Hourly Earnings and the Average for All Male Operatives (in 1934)

Jobs	Average Hourly Earnings[3]	Difference between Average Hourly Earnings and 59.5 Cents (the mean for all men in the factory)	Number of Men
Channelers..................	98.5	+39.0	3
Remainder of jobs in sole leather department.............	68.6	+9.1	11
Cutting.....................	58.3	−1.2	80
Male stitchers...............	62.5	+3.0	7
Assembling..................	37.6	−21.9	15
Goodyear stitchers...........	116.3	+56.8	7
Remainder of jobs in *making* department...............	57.8	−1.7	287
Wood-heeling...............	72.3	+12.8	45
Leather-heeling[4] (50.9 first week, 62.0 second week)....	55.8	−3.7 +2.5	18
Edge-setters and trimmers....	67.1	+7.6	18
Remainder of jobs in *finishing* department...............	56.7	−2.8	33
Treers......................	64.1	+4.6	24
Remainder of jobs for males in *packing* department........	39.6	−19.9	18

ment in the factory, displayed the least social solidarity of any department. It was composed of distinct and mutually un-

3. These figures were the average of two weeks' pay taken at an interval of three months. We transposed weekly pay and hours worked into hourly pay for these periods.
4. For the first week the leather-heelers had only nineteen hours' work. The second week was presumably more nearly representative when they then had on the average more than forty hours of work and probably speeded up their production to normal.

friendly cliques. This was particularly noticeable during the lunch hour when members of the various ethnic groups gathered in little exclusive knots in various parts of the large room. The Greeks gathered at one end of the room around an individual who read aloud from a Greek newspaper. At the opposite end of the room a group of Riverbrookers conversed. The antagonism between these two groups was particularly marked. The French and Italians gathered separately in the middle of the room, and ethnics of other nationalities were scattered about the room. In this large department the effective solidarity of each ethnic group, and the concomitant opposition and antagonism between groups greatly lessened the effective bargaining power of the makers as a whole. This lack of departmental solidarity probably accounted in part for the fact that the makers' earnings averaged 1.7 cents per hour below the factory average for men.

What we have said about worker solidarity has implied that ethnic identification accompanied a high degree of solidarity, and ethnic diversity, a low degree of solidarity. This was in fact the case, and it was recognized by workers and members of the management hierarchy. A foreman, for instance, observed: "Among the men the national tie is the strongest. For example, if I want to take on a certain man whom I have had working in the department previously, all I have to do is ask one of the workers of the same nationality and he will bring him in the next morning. Each of these groups have their meeting place and can find a particular individual very easily. Also, if I want a new man of a particular national group, I merely ask one of that group to bring a man, and he will invariably bring in one of his own nationality." The ethnic basis for worker solidarity was not so characteristic of the women operatives as of the men. The same foreman observed on this score: "The women aren't so strong on that [ethnic cliques]. They seem to form groups by proximity of working locations. Others, particularly those coming from a distance, come to work in one car and form a group, but they do not show the same grouping on the basis of nationality as the men do."

Lack of social solidarity in ethnic cliques undoubtedly accounts in part for the low earnings of women operatives as compared with those of men operatives.

In the course of studying the operations and personnel in the making room we encountered a situation in which ethnic antagonisms were partly subordinated to other considerations. The makers worked in teams of two men and each team split its team earnings. We asked several foremen to classify each of the makers for proficiency (good, medium, poor) and for speed (fast, medium, slow). We then analyzed the teams in terms of ethnicity. Our findings regarding the composition of teams with reference to each of these categories were:

90 per cent of the teams consisted of individuals of equal speed.
82 per cent of the teams consisted of individuals of equal proficiency.
63 per cent of the teams consisted of individuals of the same ethnicity.

To give us a comparison we analyzed the wood-heelers on the same three bases. In this department the men worked two at a bench, but each was paid on the basis of his individual production. Among wood-heelers we found the following correlation of characteristics for bench mates:

33 per cent of the bench mates were of equal speed.
38 per cent of the bench mates were of equal proficiency.
71 per cent of the bench mates were of the same ethnicity.

This comparison shows conclusively that when the size of the pay envelope is vitally affected ethnic prejudices are likely to be relegated to the background; when this is not the case, however, friendship and mutual interests positively influence, and ethnic prejudices negatively influence, the operative's choice of working companions.

The matter of ethnic identification has implications aside from those concerned with the social solidarity of workers. In the general value system of Yankee City, for instance, differential prestige attaches to various ethnic groups.[5] Native Yankees have the highest prestige. The Irish and the French, who have

[5] See *The Social Systems of American Ethnic Groups*, "Yankee City Series," Volume III, particularly pp. 1–102 and pp. 201–206.

been represented in Yankee City for several generations, have gained a prestige almost equal to the native Yankees—only among the elite are they still considered to be "outsiders." Newer ethnics are "foreigners" to most native Yankees, and to the Irish and French as well. They fall much lower in the scale of prestige: on the whole, the more recent the arrival of an ethnic group in Yankee City, the lower is its relative prestige. Thus, to the community, and to the workers as they represent community attitudes, the evaluation of jobs in the factory and of the individuals who hold them is influenced favorably if such individuals are Yankees, Irish, or French, and unfavorably if they are representatives of other ethnic groups. As far as ethnicity goes, the shoe factory and its jobs and workers would have the highest possible prestige if all the workers were native Yankee residents of Yankee City.

Another point of conflict existed between the workers and the community, on the one hand, and the management of the factories, on the other. The management favored ethnics who were non-residents of Yankee City. A high executive of one of the factories said: "The workers most continually on the job are the foreigners and those that live outside Yankee City. The greatest laggards are the natives of Yankee City. Many of them get aid in other ways and don't give a damn about work." This conflict in evaluations was recognized by the workers and was a sore point with them. One worker expressed it to an interviewer by observing: "You know as well as I do, and probably better, that the factory is favoring the foreigners in employment."

One of the important reasons why management prefers to hire ethnics is implicit in the statement of the factory executive just quoted. These newer ethnics are socially, as well as economically, insecure in the community. This social insecurity leads them to be more compliant, less self-assertive, than persons who are firmly rooted in Yankee City society. The newer ethnics are, therefore, more amenable than the natives, the Irish, or the French to strict social control by their bosses and to the control involved in the mechanized, set working processes. Moreover, the ethnics have little tradition of shoemaking and know little of the prestige formerly associated with the craft. Hence, they are less resistant to the mechanization of the shoe industry than are the groups which made shoes in the old days. Then, too, the newer ethnics are, because of their social insecurity, less likely to band together in effective opposition to management than are the older residents of the community. Especially is this true if the factory hires individuals of various ethnic groups, thus utilizing ethnic prejudices to reduce worker solidarities. It may also be due to the general insecurity of the newer ethnics that the factory can pay them less than it would natives, Irish, or French. This is suggested by our finding that, except for the Greeks, all the newer ethnics—the "foreigners" in the factory—averaged less in earnings than did the natives, Irish, and French. In the assembling department, where a direct comparison was possible because the pay was on a time basis, the native workers averaged 38.4 cents per hour; the combined ethnics averaged 2 cents per hour less.

3. Wage Factors

THE relation between the supply of labor and the demand for it at any given time has traditionally been regarded as a major determinant of the wages paid in various industries. In the Yankee City shoe factories we found various other factors of equal, if not greater, importance. In at least one department, however, the making department, the factor of supply and demand seemed to be an important determinant of operatives' earnings. It will be recalled that more of the ancient craftsmanship in shoe building pertains to the making operation in a turn-shoe factory than to any other job there. Makers also perform the heaviest physical labor in the factory. Yet their earnings (in 1936) averaged 1.7 cents per hour below the factory average for men. This fact may be explained largely in terms of the surplus of qualified makers. The turn method of manufacturing shoes had been traditional not only in Yankee City but throughout this section of New England. The turn method, as previously pointed out, was rapidly giving way to newer methods (in 1945 only one small factory used this method). Consequently, the number of experienced makers in Yankee City and its neighboring communities was much greater than the remaining turn-shoe factories could absorb. The competition of unemployed makers had largely dissipated the bargaining power of those still working at their trade. Hence the wages of makers were below the factory average for men.

A somewhat similar situation held for the hand cutters of outside leather (uppers of shoes). Formerly, nearly all leather for uppers was cut by hand. In 1934 machine cutting was rapidly replacing hand cutting because shoemaking was coming to be concentrated more and more in large factories where shoes are nearly always produced in large enough quantities of a given model to make it profitable to management to purchase dies and cut the leather by machine. Hence, there was a constantly growing surplus of hand cutters in Yankee City. This surplus undoubtedly explains in part our finding that the hand cutters earned less in the shoe factories of Yankee City than did the machine cutters (60.8 cents per hour average for the former, and an average of 64.3 cents per hour for the latter).

The high average earnings of the Goodyear stitchers, the channelers, and the edge-setters and trimmers may have been due in part to the operation of the principle of supply and demand. These were all jobs calling for considerable care and proficiency that had to be learned in the factory; workers could not learn these jobs except by working at the machines. This tended to restrict the supply of qualified workers. In the case of the Goodyear stitchers and channelers, however, the decline in the turn method of manufacture had presumably tended to create an oversupply of workers since these operations are performed only in turn factories. Hence some factor other than supply and demand must have operated to make these two jobs the highest paid in the entire factory. The supply-and-demand argument appears more plausible in the case of the edge-setters and trimmers because these operations must be performed in every shoe factory, regardless of method of manufacture. The secure bargaining power of these operatives, derived from this situation, was enviously recognized by other workers. A maker said once to one of our interviewers: "Don't be a maker. Be an edge-setter or a [machine] cutter. If you're a maker and they don't make turn shoes you're out of luck. If you're a cutter or an edge-setter you can always get a job."

The factor of supply and demand definitely could not account for the high earnings of wood-heelers (12.8 cents per hour above the factory average for men). The operations were hand techniques involving the use of a variety of common tools.

Some proficiency in the essential operations could therefore be attained outside the factories. Moreover, this process had been mechanized in many New England factories; therefore, the supply of hand operatives undoubtedly exceeded the demand. There seems little doubt that the social solidarity of the workers in this department destroyed the effectiveness of the relation between supply of labor and demand for it in determining wages of the operatives.

In analyzing our data on earnings, we discovered a correlation that is much more pervasive than any we could isolate for the supply and demand factor. This is the inverse relation between the number of similar jobs and the wages paid for them. This inverse relationship holds for those jobs in which the supply and demand factor might be thought dominant, and also for a great many of the technological jobs in the factories in which the supply and demand factor definitely does not determine rate of pay.

The inverse correlation between number of similar jobs and rate of pay may be clearly seen if we rearrange the wage figures in two tables: one showing the jobs for which pay was in excess of the factory average for men, and the other showing the jobs for which pay was below the factory average for men. In each case the table is arranged with the highest paid group at the top, the lowest paid at the bottom. In each case, too, the number of workers in each type of job is shown.[6]

Table 2 shows that, except for the wood-heelers and the male stitchers (other than Goodyear stitchers), there is a close inverse correlation between the number of operatives doing the various sorts of jobs and the average hourly rate of pay for the jobs. The marked social solidarity of the wood-heelers undoubtedly accounts for their displacement in this table. The male stitchers, on the other hand, are at the bottom of the list

6. In constructing these two tables (2 and 3) we have omitted the figures for leather-heelers, assemblers, and male packers other than treers. Leather-heeling was omitted because we were not sure our wage figures for this department were representative. Assemblers were omitted because their jobs had exceedingly low status—even the shoe operatives' union refused to regard assemblers as shoemakers or to admit them to the union. Male packers were omitted because, as we have said earlier, they were regarded as holding "women's jobs" and were paid on a time basis at the same rate as the women who worked in this department. We present our tables, then, with these three omissions.

because their jobs were viewed as "women's work"; that they are on the list at all is strong evidence that we are here dealing with a fundamental factor in the determination of rates of pay.

In Table 3, only the finishers are out of place in terms of the correlation we are discussing. If their earnings about

TABLE 2

Jobs for Which Operatives Are Paid More Than the Factory Average for Men

Jobs	Excess of Average Hourly Earnings over Factory Average for Men	Number of Workers
Goodyear stitchers....................	56.0	7
Channelers..........................	39.0	3
Wood-heelers........................	12.8	45
Sole-leather workers except channelers.....	9.1	11
Edge-setters and trimmers...............	7.6	18
Treers..............................	4.6	24
Male stitchers.......................	3.0	7

TABLE 3

Jobs for Which Operatives Are Paid Less than the Factory Average for Men

Jobs	Deficiency of Average Hourly Earnings from Factory Average for Men	Number of Workers
Cutters.............................	−1.2	80
Makers except Goodyear stitchers.........	−1.7	287
Finishers except edge-setters and trimmers..	−2.8	33

equaled the factory average for men, the correlation between rate of pay for similar jobs and their number would be perfect in view of other known factors. All the finishers were men, and there were no members of the newer ethnic groups. It was a foreman of this department who said that no "foreigners" would be allowed to work here. The social solidarity of the department might be expected to cause a raise in wages. No supply-and-demand factor detrimental to the interests of workers is present because the operations performed in this department were common to all shoe factories. We must admit having no ready interpretation of the low rate of pay for these thirty-three men except that the jobs, confined to polishing operations, were extremely simple. We can, however, discuss other relations between a number of similar jobs and their respective pay rates.

To summarize these two tables, we may say that of the 515 men whose jobs are compared, 400 received less pay than the factory average for men. Where such a large proportion of the total number of workers is concerned it is vital to the interests of management to keep the rates of pay as low as possible. An increase in the piece-rate in the making department, for instance, would increase the pay of 287 men and add considerably to the cost of production. In the large departments, therefore, and in any case where a large number of individuals do similar work, management benefits by keeping the rate of pay as low as possible. It is even worth the sacrifice of considerable employee good-will to make such savings. On the other hand, when the number of individuals doing similar work is small, or when a group of workers is highly integrated as in the case of the wood-heeling department, the friction and possible disruption of production that would follow reduction of piece rates are avoided by management because the overall increase in cost of production that comes from paying the higher rates is inconsiderable. In some cases, such as stamping, which was paid at a higher rate than stitching despite the simplicity of the job, management sees no important saving in the cost of production by reducing the piece-rates even though no worker opposition is expected since only two or three operations are involved.

From the workers' point of view, the fewer the individuals doing similar work, the greater their bargaining power with management is likely to be. In some cases this attitude undoubtedly arises merely from managerial inertia, as among the stampers mentioned above. In cases where a small number of workers performed key operations, however, as did the channelers and Goodyear stitchers, bargaining was on the basis of the importance of their jobs. In general, we have observed, the smaller the department or group of men performing similar operations, the greater is the social solidarity between the workers likely to be. In big departments like the making department, cliques form and disrupting oppositions and antagonisms develop among the workers which tend to destroy their unanimity in bargaining with management. In small departments like the sole-leather department, inner group oppositions do not develop to any comparable extent. Such departments are therefore able to take a stand against management to the advantage of the workers.

Another variable which correlates with earnings is the social stratification of workers. The earnings of both men and women averaged highest for operatives in the lower-middle stratum—61.7 cents per hour for men and 44.9 cents per hour for women. Men of the upper-lower class averaged 59.8 cents per hour, and women, 39.9 cents per hour. Among operatives of the lower-lower class, men averaged 57.1 and women 39.6 cents per hour. We also analyzed each department in terms of the relation between social class and earnings and found similar situations in almost all. Detailed comparison is not necessary, but we may say that the lower an individual in the social scale the less effective he is in bargaining with management. This finding would appear to corroborate the good judgment of management (from its own point of view) in discriminating against "educated" (usually lower-middle) individuals for technological jobs and to verify management's belief that individuals of the two lower classes make more docile employees.

Age has some bearing on the technological worker's earning power. For the factory as a whole, we found that average hourly earnings of men increased to the age of 39, dropped sharply in the 45-49 age group, smoothed out and then dropped again in the 55-59 year age group. Women's earnings also increased to the 35-39 age bracket, and then dropped slowly to the 50-54 age interval where they decreased sharply. The drop in the average earnings of older workers is doubtless due in large part to their loss of bodily agility, the penalty of age in the piece-rate system. In addition, management practiced preferential hiring of younger workers, a fact of which workers were generally aware. This knowledge may well have influenced some older workers who took pay cuts which younger workers would not accept.

A further factor which sometimes affects the earnings of individual workers is the fear that management will reduce the piece-rate if the worker earns too much. Sometimes the fear is common to a number of workers engaged in similar tasks and causes them all to work at somewhat less than their maximum rate of speed. In other cases pressure is brought to bear by workers on one of their number who is outstripping them at the same task. A cutter discussing this point said: "The workers gather together once in a while in the lavatory and talk about the work. If anyone says he has made four or five dollars, the others try to tell him to stop work and let them catch up." In some cases only one or two operatives work at a particular operation, and no pressure is put on them by other workers to slow down. In such cases, if the individual operator feels no fear of a rate cut and if the job is one that is capable of great speed-up, the worker may surpass in earnings many individuals engaged in more important or higher-skilled jobs. This was true of the punchers and stampers in the stitching room.

In some departments, particularly cutting and stitching, style changes have an important bearing on earnings. Hand cutters, for instance, were given the job of cutting the uppers for small runs of novelty shoes because it did not pay the factory to have dies made for the clicking machines. These cuttings were often intricate, and intrinsically more time-consuming than those for ordinary shoes. Then, too, the cutter's job changes with each style change. Hence he is unable to develop high speed on any run of novelty shoes. This is equally true of some jobs in the stitching department, particularly fancy stitching. The women who worked at these jobs did decorative, largely non-utilitarian, stitching on the uppers of shoes. The designs changed with every shift in the wind of fashion. As a result the fancy stitchers, like the hand cutters, were unable to achieve maximum speed on any one style. That these considerations have a bearing on earnings is indicated by the fact that hand cutters earned less than machine cutters, and fancy stitchers averaged less than other stitchers.

The fact that hand cutters earned less than machine cutters is another instance of disagreement between management evaluations of jobs and those of the workers and the community. The cutters were traditionally the aristocrats of the shoe trade, high community- and worker-prestige still attaching to hand cutting rather than machine cutting. This was exemplified in the statement of a former cutter who said: "I was a cutter. The cutters don't like the machine. I know how they feel. I remember when I was working on the machines. They feel now the same way I did. Every time you brought the arm down on the die, you felt like a knife was going in your back. You hated to go to work. It was a jar to your system. I used to like to go to work before the machines. It was like going to your own home. With machines it was like a knife going in your back."

Workers and the community do not always value hand work above machine work, however. It was generally agreed that the job of making was one of the less desirable ones in the factory, though it was the highest-skilled hand job left in the turn process. The low value of making in the eyes of workers and the community appeared to be due to a number of factors. Employment was uncertain because of the oversupply of makers. Wages were low for the same reason, although the work was arduous as well as exacting. Since many members of the newer ethnic groups had learned to be makers, the prestige of making had been reduced accordingly.

In contrast to making, some machine jobs enjoyed high worker- and community-prestige. Edge-setting and trimming are examples. The high worker prestige of these jobs was shown by the tendency of the edge-setters and trimmers to wear their shirts and neckties under their work aprons. Only they and the cutters did so, in both cases expressing their own high opinions of their jobs. The high value of edge-setting and trimming derived partly from the fact that these operations were among the first to be mechanized in the shoe industry and the machines had been used without important modifications for the last fifty years. These operations were thus performed mechanically in the days when shoemaking was a craft with high prestige in Yankee City. So edge-setting and trimming were traditionally associated with "the good old days" and the operatives who worked at these jobs were considered the aristocrats of the machine workers. The jobs were also valued highly because wages were good and jobs comparatively plentiful since these operations had to be performed in every shoe factory, whatever its processes of manufacture.

In summing up the evidence concerning the evaluations of hand versus machine jobs, we may say that neither management nor workers and community displayed any real consistency in their preferential evaluations. Management paid much more for wood-heeling, a hand job, than for leather-heeling, a machine job; but, on the other hand, it paid much less to the hand workers in the making room (lasters and beaters-out) than to machine workers (Goodyear stitchers), and paid hand cutters less than machine cutters. Other things being equal, the workers and the community would, on the contrary, assign higher prestige to hand jobs than to machine jobs. But the special considerations involved in the cases of the edge-trimmers and setters and the makers, for instance, obscured and blurred their evaluations. There is little doubt, however, that mechanization of the industry has done much to lower the prestige of shoe workers in their own eyes and in the eyes of the community.

We have also shown that in several specific instances management evaluations and those of workers and the community were in conflict on this basis as on many of the other bases we have discussed in this section. The values of the community are based on broader social considerations and traditional viewpoints. The aim of management is to make shoes as profitably as possible. To that end it is desirable to pay out as wages the least possible amount of money. Whenever it can do so, management ignores conflicting interests and resists control by the community or the workers that would benefit these interests. But successful business leaders are guided to a great extent by considerations of expediency; as long as the business is making a profit it is better to avoid the danger of rousing worker antagonisms to the point where production would be disrupted. Therefore, leaders in some cases compromise strict observance of the dictates of business logic in favor of worker and community evaluations to avoid friction which might endanger production. The permitted social solidarity among the workers in the wood-heeling department is a case in point. At the same time, management would be more free to follow strictly the demands of the profit-making logic if such factors as worker solidarity could be eliminated. We have seen, in this section, some of the efforts made by management to eliminate this and other disrupting factors from consideration.

VII

YANKEE CITY LOSES CONTROL OF ITS FACTORIES

1. Big City Men Take Control

TWO fundamental changes have been occurring concomitantly, in recent years, in the social organization of Yankee City shoe factories. The first is the expansion of the hierarchy upward, out of Yankee City, through the expansion of individual enterprises and the establishment by them of central offices in distant large cities. The second is the expansion of the structure outward from Yankee City through the growth of manufacturers' associations and labor unions, also with headquarters outside Yankee City and with units in many other shoemaking communities in New England and elsewhere. Both the vertical and horizontal extensions [of these developments] have gone on concurrently, each reacting upon the other. And both decrease Yankee City's control over its shoe factories by subjecting the factories, or segments of them, to more and more control exerted from outside Yankee City.

In the early days of the shoe industry, the owners and managerial staffs of the factories, as well as the operatives, were residents of Yankee City; there was no extension of the factory social structures outside of the local community. The factories were then entirely under the control of the community—not only the formal control of city ordinances and laws, but also the more pervasive informal controls of community traditions and attitudes. There were feelings of neighborliness and friendship between manager and worker and of mutual responsibilities to each other and the community that went beyond the formal employer-employee agreement.

With the vertical extension of the managerial hierarchy the social distance between the top executives, on the one hand, and the workers and community, on the other, has increased to the point where these bonds of mutual friendship have virtually disappeared. Absentee-ownership, which usually accompanies absentee-control, accentuates this condition. The conflicts of interest between the two groups of owner-manager and worker-community thus become more pronounced because they lack the bonds of mutuality. Factory policies are established in the distant offices of large concerns; neither worker nor community now has any voice in them.

The participants—manufacturers and operatives—realizing the increasing potency of these conflicts, have formed horizontal associations, aimed to protect the interests of their members from the encroachments of the opposed group. Headquarters of these associations are also outside Yankee City and their policies are determined independently of the local community. This institutionalizing of the manufacturer-operative conflicts has thus further infringed on Yankee City's control of its shoe factories and in some degree has set worker interests in opposition to community interests.

Yankee City shoe factories are no longer owned exclusively by local citizens. More and more of them are being absorbed by larger enterprises whose executive offices are in New York City. At the time of our study, the largest shoe factory in Yankee City was owned by a company which operated several other factories in New England and also owned the nation-wide ABC chain of retail shoe stores, all of which were controlled from a central office in New York. Even some of the smaller Yankee City shoe factories, although still locally owned and managed, sold most of their shoes to chain-store organizations. But before we examine in detail the organization of shoe manufacturing in Yankee City, let us see what has been happening to shoe manufacturing in general, in terms of the vertical extension of managerial hierarchies resulting from the increase in size of individual enterprises.

As a shoe manufacturing enterprise expands, either through combination with other manufacturing or distributing organizations or by self-segmentation, the business structure becomes more complex and new grades appear in the hierarchy of jobs. The division of labor in both supervisory and technological jobs is extended, and new administrative positions are created.

In the lowest grades of the manufacturing hierarchy the division of labor is carried to such an extent, as we have observed in Yankee City shoe factories, that many of the technological jobs become simple enough to be performed with very little practice by almost anyone.

The job of the factory manager is also simplified with the expansion of the enterprise. Whereas the manager of a small independent shoe factory performs two quite distinct jobs, the manager of a factory which is one of several production units in a larger enterprise has only one job. The former must concern himself not only with the internal organization and operation of the plant, but also with the whole manifold of relations between the factory and the outside world: he must be in touch with producers of materials and equipment used by his factory, with buyers of shoes, and with investors and bankers who finance the company. The manager of the local production unit of a larger enterprise, on the other hand, need concern himself only with production; the broader administrative duties are performed by a higher executive, or group of executives, at the main office of the company.

These higher executive jobs, which are created when the enterprise expands, are, nevertheless, concerned with broader problems than any faced by the manager of a small independent shoe factory. The larger concern must buy materials in greater quantities, frequently in greater variety, and from a larger number of different sources. It has more shoes to sell, perhaps of several grades, each of which must be sold to a different segment of the public. Its problems of financing are larger, requiring wider contacts with sources of financial aid. The higher executive of the large enterprise, therefore, must relate the enterprise to a much larger segment of the total society than does the manager of a small independent shoe factory. For this reason he must have very great freedom of action in performing his job.

The combination of shoe factories into larger enterprises, the creation of new superordinate jobs, the increasing freedom of action required by jobs in the higher levels of the hierarchy, and the increase in social distance between operatives and the higher executives are represented diagrammatically in Chart V.

The four small triangles represent the total personnel of four shoe factories.[1] The technological workers in each factory are at the base of the triangle, the bottom of the hierarchy. Each factory manager is at the apex of the factory hierarchy.

The large inverted triangle at the right symbolizes the range of social relations of individuals who occupy jobs at different levels in the hierarchy, with the corresponding freedom of action which must be allowed these individuals. The range of choice allowed operatives in their working behavior is exceed-

CHART V

Vertical Extension of the Managerial Hierarchy

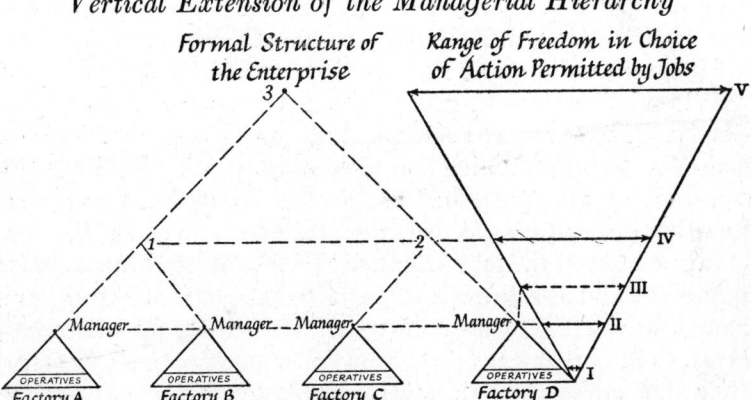

ingly small, as shown in the diagram (line I). The arrow pointing downward from this line implies the persistent tendency toward the lessening of operatives' choice of action in working relations because of increasing division of labor and mechanization. Line II in the range diagram represents the freedom of action allowed the factory manager to insure adequate internal

1. We are purposely refraining from mentioning the important divisions of a manufacturing enterprise which handle the purchase of materials and equipment, the sales of the product, the finances, and the legal division through which the rights and privileges of the enterprise are protected. These have been eliminated from our description because the typical Yankee City shoe factory of today is absentee-controlled. Almost all the activities, other than those directly concerned with production and those auxiliary activities necessary to regulate and order the production processes, such as accounting and the necessary maintenance of the building and equipment, are organized and directed from the head office in New York.

organization of the production activities of the factory. Line III symbolizes the additional freedom of action needed by the manager of a small independent shoe factory to enable him to organize the external relations of his factory, the relations with suppliers, buyers, sources of finance, the community, etc.

Now suppose shoe factory A merges with factory B, and C with D. Each plant continues operations but there are now two concerns of intermediate size where formerly there were four small ones. Two new administrative jobs are created at 1 and 2. These administrators coördinate the activities of factories AB and CD respectively, but each of the four factory managers is still responsible for the internal organization of his own plant. The factory managers, however, are not responsible for the external relations of the factory; hence their range of action is cut from line III to line II. The new executives at 1 and 2 take over the jobs of relating the enlarged enterprises to the total society. The external relations they must organize are more complex than those which had to be organized by the individual shoe-factory managers because of the increases in range and quantity of supplies needed, the size of the market that must be tapped to sell the manufactured shoes, the need for increased financing, and the growth of legal problems that accompanies growth in the size of an enterprise. Therefore, the new administrators at 1 and 2 need greater freedom of action to perform their jobs adequately than was needed by any one of the four factory managers when each took care of the external as well as the internal relations of the individual factories. This is symbolized by line IV.

If these two intermediate-sized concerns merge into one large company, still another superordinate job is created at point 3. Now the general policies of the entire large concern are dictated from 3, as are the major organizational features and production activities of all four shoe factories. The relations among all four factories, and between them and the new apex at 3, are now parts of the internal organization of the large enterprise, controlled by the administrator or administrators at 3 through the delegation of authority to individuals subordinate to them in the managerial hierarchy. Managers of individual factories are more closely than ever restricted to the job of organizing the internal relations of the factory and maintaining production. All the external relations of the enterprise, which by its increased size have now become still more complex, are ordered from 3. The executive, or group of executives, who occupies this apex of the enlarged hierarchy must therefore be allowed even greater freedom of action—as symbolized by line V—than that characteristic of jobs 1 and 2.

Theoretically, such a process of combination could continue until one huge enterprise monopolized a whole industry, or combinations of various industries. With each enlargement of an enterprise would come an expansion of the managerial hierarchy and concomitant widening of freedom of action for those at the top to enable them to cope with problems of ever-increasing complexity. Actually, however, the difficulties of internal organization usually check the growth of individual enterprises well short of complete monopoly. If these internal difficulties do not do so, the leaders of the larger society invoke antimonopoly laws so that the growth of the enterprise is stopped by outside pressure.

The ultimate limitation on the growth of the enterprise, then, shows the top executives of the enterprise to be subordinate to other leaders of the society. Even short of that point, however, the executives of large enterprises must gauge broad public opinion in setting the courses of their concerns. The head of a large shoe-manufacturing concern must predict fashion preferences, for instance, in order to make shoes the public will buy. This becomes a very large problem when a concern is so big that it must depend on nation-wide and even export sales. In many other ways, too, the top executive must understand what society demands of his enterprise and must adapt his organization to satisfy those wants. The very large manufacturing company is subject to the informal sanctions of a large part of the American public in a way that is somewhat comparable to the informal control exerted by a community on a small locally owned and managed factory.

Although the large enterprise remains, in the last analysis, subject to the control of the larger society, it tends to escape the control of individual communities in which the individual manufacturing units are located. With the expansion of the enterprise to the point where it operates several factories in different communities and has its main office and chief executives in some large city at a distance from any of the factory towns, the social distance between the top executives and the factory community becomes very great. This is true, too, of the relations between the operatives and the top executives. Every level in the managerial hierarchy above the factory manager increases the social distance between the operatives and the chief executives of the enterprise. This is symbolized, in Chart V, by the vertical distance from operatives to top executive which increases with every expansion. In large companies, therefore, the individuals at the two extremes of the hierarchy are strangers rather than friends; the top executives may issue orders in conformity with business logic which injure the interests of the workers—and not even be aware of the fact. Workers, on the other hand, do not know their ultimate boss; and, because he is a stranger, they are liable to suspect his motives and to blame him for untoward events for which he is not responsible.

Along with absentee-management usually goes absentee-ownership. In the case of a small factory, the owners as well as the manager are local residents; they belong to the community, subscribe to its beliefs and ideals, and are subject to its social control. When the local factory is but part of a large enterprise, however, the owners as well as the top executives are strangers. Frequently, ownership rests with a large number of stockholders who have no personal interest in the social structure or operations of the enterprise. In such cases the workers and the factory community suffer further loss of control over the factory through their complete loss of social control over the owners.

Yankee City at the time of our study had partly lost control of its shoe factories through the development of absentee-management and absentee-ownership such as we have described. During the early part of 1933 there were nine shoe factories operating in Yankee City, one of which was absentee-owned and managed. At one time during the progress of the study only the absentee-owned factory was in operation, and at the close of our investigations it dominated shoe manufacturing in Yankee City. The organization of shoe manufacturing in Yankee City is represented by Chart VI.

The shoe factories A, B, and C in this figure are all located in Yankee City. A and B are locally owned and independent, but factory C is owned by outside interests which also own shoe factories X and Y, located in other communities, and a chain of retail shoe stores also located outside Yankee City. Factories C, X, and Y and the chain of retail stores are all managed from one main office in New York City.

Similar manufacturing processes are used in the three Yankee City factories, and the operatives in all three factories occupy similar low levels in the business hierarchy. The managers' function as organizers of production is likewise similar in the three factories.[2] In formal internal organizations the three Yankee City shoe factories are, therefore, essentially

CHART VI

Shoe Manufacturing in Yankee City

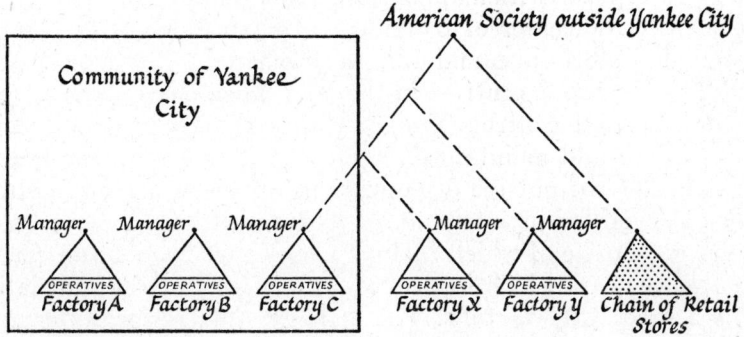

similar. The differences between the locally owned and managed factories (A and B) and the absentee-owned and managed factory (C) are to be sought in the manner in which they are related to the community and the outside world.

The managers of the locally owned, independent shoe factories A and B direct the external relations of their factories from their offices in the local community. Besides their concern with the internal organization of their factories, they must

---

2. For the sake of simplicity we are omitting discussion of the intermediate jobs of supervisory control between the managers and the operatives as these jobs need not be considered in dealing with the relations between the factories, with the community, and with the larger society; these jobs, however, are also similar in the three factories.

procure supplies of material and equipment, secure a market for their shoes, and obtain whatever financial aid the factory needs in the form of either capital or credit. The manager of the absentee-controlled and owned factory C, however, is concerned only with the internal organization of the factory he controls; all the external relations of this shoe factory are ordered from the New York office of the large enterprise of which it is but one production unit.

This arrangement gives shoe factory C a distinct business advantage over the locally owned factories A and B. This is because vertical extensions comparable to those we have described for the shoe industry have occurred also in the case of many of the organizations with which the head of a shoe manufacturing enterprise must maintain close relations. Many of the producers of the materials and equipment needed in the manufacture of shoes have expanded like the shoe business into large enterprises with main offices in New York City. New York is likewise the center of banking and finance. Many chains of retail shoe stores also maintain head offices in New York.

Thus the top executives in the New York office of the large enterprise that controls Yankee City shoe factory C are able to establish and maintain close social relations with the heads of many of the outside organizations on which a shoe factory must depend for maintenance of its production. Moreover, the large enterprise of which factory C is a part has its own chain of retail stores, which insures an outlet for the shoes made in this Yankee City factory. The managers of the locally owned shoe factories A and B, on the contrary, cannot establish these close social relations with outside organizations. They must spend most of their time in the factory office and establish outside contacts by mail or wire; only occasionally can they go to New York or elsewhere to establish personal relations with organizations whose coöperation is essential to the business welfare of their factories. In comparison with factory C, the locally owned factories also are greatly handicapped because, having no retail chains of their own, they do not command the certain market for their products that factory C does. All of these advantages enable shoe factory C to maintain more continuous production closer to its capacity, making it a more efficient production unit than either of the locally owned and managed factories A and B.

The factor of vertical extension and absentee control which is so advantageous to factory C in a business way puts it at a disadvantage, compared with factories A and B, in insuring smooth functioning of the internal organization of the factory. In locally owned factories A and B, an operative knows his ultimate boss. In many cases operatives and managers have been acquaintances, perhaps even friends, all their lives. This induces close social relations between workers and management, mutual trust, and some recognition by each of the point of view and problems of the other. When situations arise wherein the interests of the two groups diverge, some compromise is likely to be effected after essentially friendly discussions between workers and management. In the case of factory C, however, the social distance between the operatives and the top executives who live and work in New York City is tremendously increased. Neither knows nor understands the other. Operatives' relations with the owners—stockholders, in this case—are even more distant.

A result of all this is that operatives are suspicious of the motives of strangers, and the higher executives and owners fail to appreciate the social needs of operatives. Small dissatisfactions of operatives are thus likely to be greatly magnified. They have, moreover, a sense of insecurity in their jobs which is not overcome by the greater regularity of employment offered by factory C as compared with factories A and B. Workers in factory C cannot feel as sure as those in A and B that the factory is an integral part of the community of Yankee City. Operations of factory C might be removed to another community at the whim of the unknown persons in ultimate control of it. This is not an idle fear. Shoe-manufacturing operations can easily be moved from one community to another because factory buildings and much of the machinery are rented rather than owned by the manufacturer. And many communities in New England offer inducements, besides an adequate supply of labor, equal to or even superior to those advertised by Yankee City. There have been numerous instances in recent years of such movements of shoe-manufacturing operations from one New England community to another or even outside New England altogether. Thus the maintenance of a smoothly functioning internal organization is more difficult for the manager of factory C than it is for either factory A or B because of the great social distance between the top and bottom of the hierarchy of which this shoe factory is a part. This social distance, in turn, is the result of the vertical extension of the managerial hierarchy through the growth of the enterprise of which factory C is a part.

The vertical extension of its managerial hierarchy and the absentee control of factory C also made its relations with the community of Yankee City much different from those of factories A and B. Typically, the managers and supervisory staffs of the locally owned, independent factories (like A and B) are native Yankees; frequently, they were born and grew up in Yankee City. They are thus involved in the general social life of the community, belonging to various associations, clubs, and other organizations. They subscribe to many general community attitudes which impinge upon their working relations, frequently causing them to modify working behavior away from that which would follow the single-minded dictates of the profit-making logic. Their business status and, especially, their general social status depend upon the opinions of their fellow townsmen. Their life in the factory is not divorced from life in the community outside of working hours. Part of the motivation that determines their business behavior is the desire that fellow townsmen regard them as upright and fair business men who treat their employees properly. Such desires frequently militate against their acting strictly in accordance with the profit-making logic; business advantages are sometimes sacrificed because the manager (or owner) places greater value on community prestige than he does on increasing factory earnings by some means which would endanger that prestige.

In the case of absentee-controlled shoe factory C, however, there is little business need for the manager to participate in community activities. He can operate the factory in strict accordance with the orders of the main office more easily, in fact, if he does not take part, either as an individual or as a representative of the factory, in local associational activities. If he does involve the factory in community activities, he involves it and the larger enterprise of which it is a part in community responsibilities and subjects it to community pressure and control. This is precisely what the top officials of the large enterprise do not want. They want the factory to be as free as possible of community pressures so that its operations can be dictated in strict accordance with the profit-making logic. This would even allow them, if it seemed desirable, to move their factory to another community.

In view of the advantages that accrue to the total enterprise if factory C is kept free of informal community pressure and control, there appears to be special significance in certain differences between the persons who hold similar jobs in locally owned factories A and B. In factory C the manager and many of the subordinate members of the supervisory staff reside outside of Yankee City. Many of them are also members of ethnic groups that have comparatively low prestige in Yankee City. This may be merely a coincidence, but it seems more likely that it is explicit company policy to hire supervisors who are not firmly entrenched in the community life of Yankee City.

However this situation came to be, its effect on the community and workers of Yankee City has been to increase the social distance between them and the local managers of factory C. It has also contributed, along with the division of labor and mechanization, to lower the prestige of shoemaking as an occupation in Yankee City; for people do not like to take orders from those they regard as social inferiors—and many of the native Yankees, Irish, and French who are the traditional craftsmen of the shoe trade in Yankee City do regard members of the present supervisory staff of factory C as social inferiors. This attitude tends to precipitate the withdrawal of such persons from the industry; and they are replaced largely by members of the newer ethnic groups, again lowering the prestige of shoe manufacturing. Loss of prestige of the industry means some further loss of effective control of its shoe factories by the community since societies protect more diligently the forms of institutions having high social value than those whose value is considered low.

As far as Yankee City shoe factory C is concerned, the manufacturing processes have become so standardized, and personal relations and community control so minimized, that the production of shoes requires little time or attention from the higher executives. These men can thus concentrate on the problems of integrating the enterprise into the larger society. The problem of merchandising, for example, has absorbed much of the working time of these higher executives, but this essential operation is coming to be standardized, too: witness the development of the chain-store method of selling. With both production and merchandising largely routinized, top executives devote most of their attention to other problems of widening the extent of the enterprise's relations with the total society. The financing of the large and expanding enterprise, for instance, assumes major importance.

Other even broader problems are looming on the business horizon. "Institutional advertising," for instance, which has increased greatly in recent years, consists of efforts to "sell" the American public on the virtues of an individual company or even of a whole industry rather than merely on its products. Such attempts to expand the base of integration of business enterprises in the total society were hardly considered to be business problems at all a few years ago. It is to be expected that, as individual businesses grow larger and affect the lives of ever-increasing segments of the American public, these very generalized problems of establishing satisfactory relations between the businesses and the total public will occupy more and more of the time and effort of the top executives.

Small, locally owned and managed shoe factories, like Yankee City's factories A and B, face powerful competition today from shoe factories like C, which are production units of large enterprises. We have mentioned some of the reasons why this is so, but even before the development of such large enterprises there was a high rate of mortality for Yankee City shoe factories. Over forty factories have failed in Yankee City in the past seventy years. Today the difficulties of these small independent shoe factories are greater than ever before. They find it increasingly difficult to so organize their external relations that they can insure sufficient volume and continuity of production to meet the price competition of the larger and more efficient enterprises. It seems reasonable to expect that large enterprises, like that with which factory C is associated, will become increasingly dominant in New England shoe manufacturing.

This trend, together with the type of integration of external relations characteristic of these large enterprises which we have described above, indicates the sort of role towns like Yankee City must prepare themselves to play in American industrial life. Yankee City needs shoe factories to absorb its plentiful supply of efficient labor, for it is highly important to the well-being of any community that its labor be gainfully employed. Factories like C, absentee-controlled, can best assure local labor of continuous employment because they operate more constantly than small independent plants and also because the large concerns do not suffer as high a rate of mortality.

Such factories, as we have shown, resist many forms of social control by the local community, particularly the informal controls. If the community seeks to impose control, the absentee directors are likely to remove operations from the community. In order to keep such factories, Yankee City must accept a subordinate position and cater to their needs rather than try to control them as the community once controlled the small locally owned factories. Local pride resists the acceptance of this subordinate role as it decries absentee control and ownership. But in the end, if Yankee City is to prosper under the present system of industrial organization, it must learn to play the part it is equipped to play in the development of the industry, even though, for the advantages it will gain thereby, it must relinquish some of its cherished independence.

The full significance of this elaboration of the vertical hierarchy and its powerful effect on the instigation and outcome of the strike will be treated in the chapter which follows; the effects of horizontal extension of the shoe factories are treated in the next section.

2. Horizontal Extensions of the Factories

ALONG with the vertical extension of the shoe factories has gone another type of extension outside the factory community, which we are calling the horizontal extension. It consists of the organization of grades in the manufacturing hierarchy across factory and community lines into huge associations, the scope of some of which is industry-wide. Such associations are organized primarily to protect the interests of one grade in the manufacturing hierarchy against the encroachments of other grades. The rapid development in recent years of associations of owners

122 *The Social System of the Modern Factory*

and managers, on the one hand, and of workers, on the other, has gone on concomitantly with the growth of extreme vertical extensions in the shoe business. Manufacturers' associations have been formed without public resistance, but labor unions have been resisted both by management and by some sectors of the general public. This dual development reflects the increasing seriousness of the conflict of interests between the different grades which has accompanied the increasing social distance between them. As far as communities like Yankee City are concerned, these associations further decrease the community control over the shoe factories because the headquarters of the groups controlling the Yankee City units are outside the com-

CHART VII

Horizontal Organization in Shoe Factories

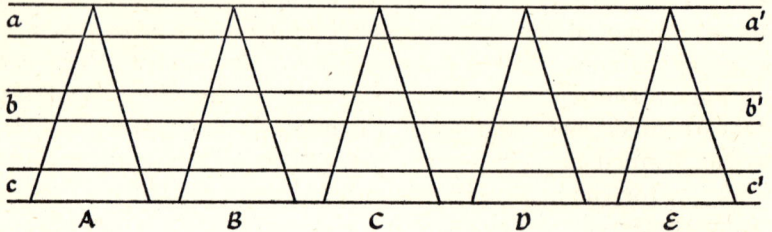

munity and independent of it. These horizontal associations, like the vertical extension of managerial hierarchies, represent integrations of the shoe industry into the larger society, automatically decreasing their integration with the local community.

In portraying the associations which reflect the horizontal extension of relations, we will consider each enterprise as a total manufacturing system, without discriminating between those composed of several units and those having only one: triangles A, B, C, D, and E (in Chart VII) represent not individual factories but whole enterprises. Bar aa' cutting across the tops of the triangles represents manufacturers' associations as a group; bar bb' represents employers' associations. An independent owner operating a single plant might be a member of both a manufacturers' and an employers' association. Among enterprises composed of several manufacturing units, the various factory managers may be members of employers' associations. The lowest bar, cc', represents the labor unions, which have the greatest number of people to organize.

Horizontal organizations consist of associations formed to protect the working statuses of their individual members. The latter are bound together by reason of their mutuality of interests within the broad industrial hierarchy. Recognition of this mutuality of interests creates a solidarity which, when organized, results in a greater bargaining power for the members of any of these associations within a given factory or a whole industry. Implicit in the logic of such organizations is recognition of the fact that the interrelations in an industry are not comprised solely of the vertical relations of subordination and superordination.

Protective associations are motivated by a logic more abstract than the profit-making logic, and they do not function at all times with the same intensity. The intensity varies with the imminence of the menace which they are organized to resist; in times of crisis, it becomes very great. The purpose of any one of these associations is the protection of the interests of the totality of its own membership, irrespective of factory boundaries. Nevertheless, in specific instances of localized antagonisms and tensions, the functioning of these associations focuses on intra-factory relations.

Just as the structure of a business enterprise is subdivided for efficient functioning, so each of these organizations has a hierarchical structure. The national headquarters is usually found at the center of the shoe industry's operations; the sub-organizations are situated at various points throughout the area in which shoes are manufactured.

There are few organizations representing any one grade which function universally throughout the shoe industry. Conflicts arise among different organizations formed to protect the interests of a certain grade. This is particularly noticeable in the conflict between the American Federation of Labor and the Congress of Industrial Organization. We shall have more to say about this a little later.

Inasmuch as employers' associations are generally organized on a smaller scale than the other two types and are more or less localized, this section will consider only two classes of protective associations: manufacturers' associations and labor unions.

Manufacturers' associations are established to further the interests of the superordinate executives of the various enterprises in the shoe industry. They are simpler to organize than labor unions because, compared to the workers, the number of individuals functioning in the role of merchant-manufacturer is small. Their specific purpose is to enable the executives thus organized (1) to deal more adequately with the problems involved in relating the factories to the larger world, e.g., problems of vertical extension; and (2) to protect themselves as a group against other horizontal associations organized in the interests of other groups.

The current national organization of shoe manufacturers is known as the National Boot and Shoe Manufacturers' Association.[3] Some indication of the aims of this organization to control policies in the shoe industry can be seen from the aggressive role it played in the formation of the NIRA code for the industry. The initiative for the development of a code for the shoe industry was taken by the National Boot and Shoe Manufacturers' Association. It foresaw the enactment of some form of legislation prior to the passage of the National Industrial Recovery Act, and accordingly attempted to bring the shoe manufacturers together to obtain consensus ~~of opinion~~ on what should go into a code for the industry.

After the code was formulated, the N.B.S.M.A., in conjunction with the Administrator of the NIRA, became the agency for administering its provisions. The Code was presented to the President on October 3, 1933. Between that date and the date of the adverse Supreme Court decision, May 27, 1935, a series of hearings and conferences was held. The NIRA marked a new development in dealing with the problem of competition. Its creation was a recognition that all industry required coöperation within and among manufacturing groups.

---

3. There are also a number of more or less local manufacturers' organizations which need not be discussed here. After we have discussed the development of labor unions, we shall compare manufacturers' associations with labor unions in the shoe industry.

Yankee City Loses Control of Its Factories

We have mentioned previously the apparent lack of divergence of interests between merchant-master and journeyman in the earlier days of the industry. It appeared openly for the first time about 1868 in the labor organization known as the Knights of St. Crispin. This short-lived organization was followed, roughly a generation later, by the National Boot and Shoe Workers' Union. The union was highly centralized, and, in order to create a more democratic form of organization, the Shoe Workers' Protective Union was formed a few years later. Both these organizations still exist. The latter, in fact, called

CHART VIII

The Change from Craft to Industrial Unionism and Its Relation to Division of Labor and Destruction of the Skill Hierarchy in Technological Jobs

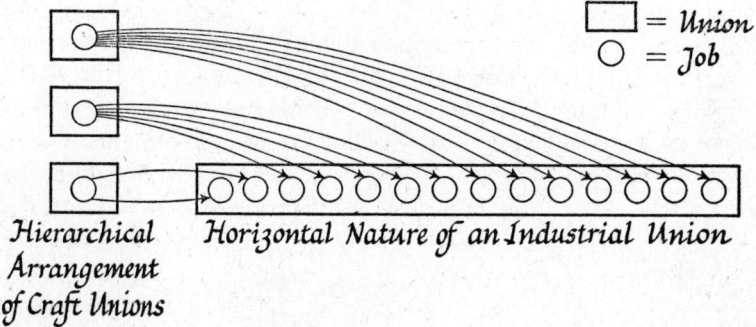

the shoe strike of 1933 in Yankee City, but it decreased in power thereafter. The National Boot and Shoe Workers' Union, as its name implies, is a national organization. It has continued to be an important force in many centers of shoe manufacture.

Early in 1934, the United Shoe and Leather Workers' Union absorbed the two locals of the Yankee City Shoe Workers' Protective Union and started to organize the workers under a broader and more extensive national program. In 1936 the U.S.L.W.U. was the dominant union in Yankee City. It is an industrial union to which all shoe operatives—regardless of sex, race, or creed—belong. The Shoe Workers' Protective Union has retained a small and relatively unimportant foothold in Yankee City among certain specialized groups of workers.

The purpose of the U.S.L.W.U. is shown by the following excerpt from the preamble to its constitution:

1. The UNITED SHOE AND LEATHER WORKERS' UNION has been organized to enable all workers engaged in the shoe industry and all component parts thereof, regardless of creed, color, sex, nationality, political or religious affiliations, to unite under one banner for the better protection of their economic interests and for the betterment of their working and living conditions.

2. We hold that against the organized forces of the bosses in the industrial field, the workers must, in order to protect their interests, build and develop powerful labor unions so organized as to have the controlling forces where the workers are employed.

3. We affirm that the form which the labor union must take must correspond with the prevailing form of the organization of the industry, and the labor union must be prepared to continually adapt its form, policy and tactics to correspond with the development of the form of organization and tendencies of that industry.

4. In order to attain our objective, we, the UNITED SHOE AND LEATHER WORKERS' UNION, do declare ourselves in solidarity with the rest of the working class who are striving to do away with wage slavery and for the purpose of securing for the workers an ever increasing share of the material goods of the world; of the intellectual and spiritual attainments of civilization and of the fruits and jobs of life, and ask all wage earners engaged in the shoe industry and the component parts thereof to unite with us under the following constitution.

The object of the U.S.L.W.U. as expressed in Section 4 of Article I in the Constitution of the Union is as follows:

The object of the U.S.L.W.U. shall be to obtain and preserve for all workers engaged in the shoe industry and its component parts just and reasonable conditions of work with respect to wages, working hours and other conditions of employment; to secure sanitary surroundings in their places of work and humane treatment from their employers; to aid needy workers in the industry; to cultivate friendly relations between them; and generally to improve their material and intellectual standards. Such objects shall be accomplished through concerted efforts to organize the unorganized workers in all branches of the industry; through a militant program of strikes; through negotiations and collective agreements with employers; through the dissemination of knowledge by means of publicity and lecture courses and through all other means and methods usually employed by organized workers to maintain or better their standards of life.

The aims and purposes of this industrial union, as described in the foregoing excerpts from its constitution, are highly significant if considered in conjunction with the division of labor and mechanization which have occurred in the shoe industry as described in Chapter V. We showed there that the technological hierarchy which once existed in the shoe trade has now almost vanished from the shoe factories. The early unions in the shoe industry were craft unions with a hierarchical relation to one another similar to the hierarchical relations of the various technological levels of the workers they organized. The U.S.L.W.U., with its emphasis on the quality and common interests of all shoe workers, overtly recognizes that the former technological hierarchy has indeed broken down in the shoe industry. The fact that the union grew so rapidly in size and importance during the early 'thirties shows that shoe workers themselves were aware of the breakdown. A comparison of the two kinds of unions, craft and industrial, is represented by Chart VIII, an adaptation of Chart IV which showed the breakdown of the skill hierarchy in technological jobs.

Labor's great internecine struggle, which we have witnessed in the struggle between the A.F. of L. and the C.I.O., can be understood in terms of our analysis of the break in the skill hierarchy if that analysis is given wider application than we have thus far attempted. The A.F. of L. strives to maintain the ideal of hierarchies in technological jobs, arguing that the workers at different levels in such hierarchies have different interests from other workers. The C.I.O. argues that technological jobs in a large segment of modern industry have been reduced to a common low level of skill and that the A.F. of L.'s doctrine is antiquated and non-functional in situations like that of the Yankee City shoe industry today. C.I.O. thus wants to organize all the technological workers in a given industry on a basis of equality in order to fight management and owners in the common interests of labor. What the outcome of this struggle within labor will be, no man knows. It appears, however, that each kind of organization fills a need in various realms of American industrial life. The technological jobs in many highly mechanized industries are probably all, or nearly all, on a common low level of skill comparable to that required in the modern shoe industry. In these enterprises, the industrial type of union is functionally consistent with both the actual interests of the workers and their relations with management. There is no doubt, on the other hand, that in some industries there is still a definite hierarchy of technological jobs. In such industries the traditional craft unions probably are more consistent, functionally, with the working relations and interests of workers than the industrial type of union.[4]

The local unions of the U.S.L.W.U. in Yankee City have a central executive board to which the union members annually elect individuals from the various production departments of the factories. The officers of a local include a president, vice-president, secretary-treasurer, business agent, and members of the executive board. The business agent is the officer who represents the local in major negotiations with the factory. Minor union problems within the factory are in the hands of "stewards," elected by shop crews in the production departments. The funds of a local are under the control of trustees who are responsible to the executive board.

The locals are closely affiliated but have separate organizations with separate agents. Workers in the cutting and stitching departments of the factories are organized in one of the

---

4. The very terminology of unionism, although it is confusing, betrays the conflicting ideologies of the two types. Craft unions are called "horizontal" unions, the implication being that they organize separately the workers at various levels in a technological hierarchy. Industrial unions are called "vertical" unions, implying that they organize in one union workers at all levels of a technological hierarchy. The hierarchy is so taken for granted that it is implied in describing the type of unionism that developed because of the breakdown of the hierarchy.

locals, and workers in all other departments in the other. This division of union activities under two locals reflects merely an expedient organization device; cutting and stitching rooms are always adjacent and the operatives of these departments, who form a large group including both sexes, all work on the same sorts of materials. The division in union activities does not represent any differential in prestige between cutters and stitchers and other operatives.

Both locals of the U.S.L.W.U. in Yankee City have the same headquarters in a building near the center of the town. This building has a central room for union meetings and two small rooms which are the offices of the two locals. From time to time, lectures on subjects interesting to the workers are given in the main room or union hall. The headquarters is a type of workers' club; it functions as a social meeting place according to the amount of, and interest in, union activities.

The executive boards of both locals of the U.S.L.W.U. at the date of this study (1935) consisted exclusively of native Yankees, Irish, and French. These boards, representing the important departments of the productive division of the factory, were elected by the total membership. The members of the executive board served without pay. The choice of members reflects the prestige which was enjoyed by Yankees, Irish, and French over those ethnic groups which had more recently arrived in Yankee City. Thus we see the phenomenon of an association which is founded on the technological logic that all workers are equal denying that logic because of the existence of social classes and a system of sentiments associated with these classes in the community.

The establishment of the United Shoe and Leather Workers' Union reflected a demand for the organization of workers in the shoe trade under a logic which had as its base the common interest of all workers in the industry. This union made a strong appeal because of its extremely democratic organization in which the ultimate control lay in the hands of the members. The officers were elected for comparatively short periods and were subject to recall at any time.

Although the original appeal of the U.S.L.W.U. was effective in expanding its influence, it failed to hold the workers' interest. It expanded rapidly: by September 1935, its executive secretary estimated its total membership at 60,000, claiming that it then represented the largest union membership in the shoe industry. As conditions improved in the industry, however, the antagonisms between workers and those in supervisory jobs were reduced to minor grievances, comparatively easy to arbitrate by the executive board of a local, if not by the business agent or stewards. As a result, the influence of the controlling ideology of the union over the members lessened, and attendance at meetings of the locals was reduced.

By the end of 1935, only one large factory in Yankee City—whose operatives were organized under the U.S.L.W.U.—was functioning steadily. Soon after that time, however, one of the formerly important factories was reorganized and started production on a small scale. The union was not strong enough to demand recognition in this smaller establishment. Consequently, by the spring of 1936 the latter was operating as an open shop, leaving only the workers of the one large absentee-controlled factory affiliated with the union. This shows that the workers are apparently unable to maintain an effective organization to protect their interests in the employer-employee conflict of interests in a slack period.

Even while the interest of the workers in the activities of the U.S.L.W.U. was waning by reason of the improving employment conditions in the industry, the union did perform an important function in demanding adherence to regular working hours, the maintenance of established rates of pay, and an equalization of the pay of different workers in the slacker periods. The union has rectified several long-standing complaints of the workers. An interviewer asked a shoemaker what the factory did when work ran low. He answered: "The night before they put out the stock to be worked on the next day, so you know how much work you have to do the next day—that way you know there's only enough for a half day's work, and you come, and when you get through, you go home. You don't have to wait around when you're not working, and if there is no work for the next day you know it the night before, so you don't go to the factory at all." The interviewer remarked that formerly there had been complaints to the effect that the men spent a day at the factory and worked for only two or three hours. The operative said: "Yes, that was before the union, but now the union won't let them do that. Before the union came, some factories would work this same way. They would let you know the night before, but a lot of them didn't; you had to come every morning and stay all day whether you worked or not. But now it's different, and it's much better this way."

The evolution of manufacturers' associations and labor unions gives evidence of the extreme complexity of the relations in the shoe industry which transcend the limited set of relations formed within any single enterprise. Their formation further evidences the basic conflict of interests between those in positions of control and the shoe operatives. Because the superordinate executives have many problems of a highly individualized nature to solve, it is somewhat difficult for them to effect a closely knit protective organization. This difficulty is to some extent offset, however, by the fact that the common menaces and problems which such organizations are established to combat are sufficiently pressing at all times to keep the associations active. Thus the problem of bridging the gap between crises is not so great as that which faces the labor unions.

By reason of the character of the relations organized by manufacturers' associations, their appeal is based upon a more concrete need than that of the labor unions. The former are established primarily to protect the business interests of the owners, whereas the latter are established to protect something less definite, the rights of the workers. The shoe strike illustrated this point. Most of the formal demands of the strikers concerned wages and the recognition of the union, but interviews with workers during and before the strike clearly showed that many of the basic grounds for dissension had little to do with the amount of wages received. The workers believed that these basic grounds for dissension among different individuals could be gradually adjusted if the union were recognized. The union served as a composite symbol of protection for all those social values which were in themselves unexpressible as well as for the protection of reasonable wages. To the worker, the union represented a medium through which his work within the factory would be better integrated with his life outside the factory.

Since manufacturers' associations purport to protect business rights of the industry as a whole, it is comparatively simple to keep them well financed by contributions from the enterprises represented. Labor unions, on the other hand, comprising a large number of individuals with little personal wealth, find it difficult to maintain the interest of the workers while the latter are reasonably satisfied with working conditions. The manufacturers' associations are usually supported by annual dues whose amount is not burdensome nor constantly noticeable. The weekly dues of the labor unions are a continual drain on the pocketbooks of individual workers with little money to spare.

When the sentiments of opposition or antagonism between operatives and organized leadership in the shoe industry become intense, the feeling of solidarity among operatives in various factories is greatly strengthened. During these times, the operatives have an active interest in their collective welfare; and it is then that the labor unions become the instrument by which this solidarity is made an effective force in negotiation with employers. When the factory work progresses in an orderly fashion, the operatives are apathetic about collective representation. During such times, the labor unions carry on a variety of minor activities in the interest of the workers which require the maintenance of local union headquarters near the factories. But in the periods between employer-employee crises the interest of the workers flags, collection of dues falls off, and the union has difficulty maintaining its organization.

With reference to Yankee City, the significance of the development of national manufacturers' associations lies in their connecting up of local industrial problems with those of the greater industry. Many of Yankee City's industrial problems are not considered in Yankee City but at associational headquarters elsewhere. This same dependence on organizations located in larger cities is evident in the administration of labor unions. Labor unions, considered as another example of the horizontal extension, are organized primarily to protect the welfare of the shoe operatives as individuals and to regulate the interrelations of the workers as a class with both manufacturers and employers. Thus, to the Yankee City community, a very important result of both the horizontal and vertical extension of industry is to lessen its independence and freedom of choice in meeting problems having to do with the livelihood of its citizens. Thousands of other small industrial cities are also involved in the process of wider integration occurring throughout American industry. They, too, must come to accept the roles in this development that are assigned them by the larger society.

Before discussing the full meaning of the vertical and horizontal extensions of the shoe business, we will examine what these extensions mean in terms of personalities by comparing present and past owners of Yankee City shoe factories.

VIII

MANAGERS AND OWNERS, THEN AND NOW

1. The Managers of Men Were Gods

THREE dead men played powerful, important, and, at times, decisive roles in the outcome of the strike. Paradoxically, although they were former owners and managers of the factories, their influence materially aided the strikers and helped defeat management. Throughout the struggle, the owners,[1] workers, and most of Yankee City continued to recognize the great wisdom of these dead owners and managers and always bowed to their judgments. The authority of these men accordingly was constantly quoted by each side to gain approval for what it said and did and to stigmatize the words and actions of its antagonists. The peacemakers quoted the deeds and sayings of the three as parables and precepts to force the warring parties to come to agreement. It is unlikely that the actual behavior of these three men corresponded to the symbols into which they had been fabricated by those who remembered them after their deaths. But it is certain that the values inherent in them as collective representations ordered and controlled much of the thinking of everyone and greatly contributed to the workers' winning the strike.

The three—Caleb Choate, Godfrey Weatherby, and William Pierce—were constantly quoted; episodes from their exploits, as brilliant Yankee City industrialists and wise and generous

---

1. Each person, each institution, and each incident is a composite drawing. As we said in Volume I: "No one actual individual or family in Yankee City is depicted, rather the lives of *several individuals* are compressed into that of one fictive person.... The justification for these changes lies in our attempt to protect our subjects and to tell our story economically. We have not hesitated to exclude all material which might identify specific persons in the community; and we have included generalized material wherever necessary to prevent recognition. The people and situations in some of the sketches are entirely imaginary. In all cases where changes were introduced in the reworking of our field notes, we first satisfied ourselves that they would not destroy the essential social reality of the points of the original interview. Only then were such materials included in our text."

employers of Yankee City men, were frequently spoken of and applied to present conditions in the shoe industry to the detriment of the contemporary managers and owners. Since the sagacity of the three verged on the supernatural, no flesh and blood owner living in Yankee City at the time of the strike could hope to measure up to the standards of these demigods. It is small wonder that managers felt weak and inadequate when they compared themselves with the great men of the past, and it is certain from their utterances and deeds that they shared feelings of guilt in the presence of their accusing employees. Their private knowledge of themselves and faith in the great managers of the past made them weak, for now that myth-making had done its work and mortals were translated into gods, the prosaic men of the present could never hope to compete with these heroes and demigods who plagued them from the past.

We will examine the evidence to see what these men were in real life but only briefly since it was what the men and women of the strike believed them to have been that made them important for this study. In the section which follows we will discuss the social personalities of the contemporary managers; then we will compare the evidence from the past and present to learn why the three dead owners were still powerful when their successors, with all the recognized glory of modern technology to support them, were considered weak and inadequate.

Caleb Choate was a pioneer in large-scale shoe manufacturing in Yankee City; both Godfrey Weatherby and William Pierce received their training under him.

Mr. Choate's success is indicated by the fact that from a capital of $100 (in 1866) he had built up a business with annual gross sales of a million dollars. By 1877, at the age of thirty, he was the "head of a large and successful manufacturing business and was one of the solid and respected citizens of Yankee City. . . . A large factor in the early success of the business was the prompt adoption of the McKay stitching machine while other shoe manufacturers were considering whether it would pay. Mr. Choate was one of the first to combine the many parts of shoe manufacturing under one roof and to successfully operate a large establishment where from the raw materials shoes were made up complete from start to finish under the management and care of one man. By 1892 annual sales totalled $1,440,250."

Before Godfrey Weatherby died he said: "There was almost entire absence of bitter feeling between Mr. Choate and his employees. . . . There were no strikes by reason of dissatisfaction with wages. He commanded respect rather than won popularity. He was a kindly man and just, intolerant of inefficiency and dishonesty, always master in his own sphere, a good judge of men. Leaders among the workmen were satisfied that they were justly and fairly treated."

Another decided factor in Caleb Choate's business success was the financial help given him by Mr. Davis Cole, a member of an "old Yankee City family." Mr. Cole was in business in Boston and retired at the age of forty-four with a fortune. He owned a home in Yankee City and planned to live on the income from his savings, placed in the bank at 5 per cent interest. Mr. Choate, with much ambition and very little money, started his shoe concern and asked Mr. Cole to endorse some notes. Mr. Cole agreed to this several times, so the story goes, without paying much attention to the amounts. Three years later, Mr. Cole was called over to the bank and learned that he had endorsed $100,000 worth of paper. Caleb Choate said "if Cole put him into bankruptcy he wouldn't get ten cents on a dollar but if he was allowed to continue he would get one hundred cents." Mr. Cole, who had complete confidence in Mr. Choate, was put in active charge by the bank and the business was successful.

Reference is made to this venture in Mr. Choate's memorial volume: "To Mr. Cole he (Caleb Choate) always expressed a grateful sense of obligation for the courage which he manifested and confidence he placed in the ability and integrity of the managing partner."

According to an elderly Yankee City business man (upper-upper), who at one time was in partnership with him, Caleb Choate employed ten salesmen who visited the "whole of the United States from the Atlantic to the Pacific. Up to 1900 he manufactured many times more than all the other Yankee City manufacturers put together."

"His business methods were exceptionally fair and praiseworthy," another said. "The fact that he was in any way connected with an enterprise was all that was needed to inspire complete confidence. His name stood for quality, honest value, fair treatment and good service in almost every city in the United States."

Caleb Choate started Godfrey Weatherby in business. According to Godfrey Weatherby's friends, they were very different. Mr. Choate was decided and abrupt, while Godfrey was more gentle and kindly. "He wouldn't be taken in by anyone," an old lady from Hill Street said, "but if a man showed he wanted to do what was right Godfrey always stood ready to help him. He became very successful and always showed a great interest in community affairs. Everyone trusted him and he never had any labor troubles of any kind. He had been to the local school with a good many of his employees and they knew his word was as good as his bond."

According to the most important and respected opinion-maker in the city, "if he [Weatherby] had been in business today he wouldn't have had any strike because his employees believed in him and he would have put all his cards on the table."

One of the partners of a large firm which still bore the name of Weatherby said: "Mr. Weatherby has been dead a good many years but we kept his name because he was such a fine man and his name meant so much. He did more for Yankee City than anyone else. If it hadn't been for Mr. Weatherby many big Yankee City companies would have been on the rocks."

At the time of the strike another upper-class informant said: "Mr. Weatherby was highly revered and respected and there's scarcely a meeting during this strike where his name isn't mentioned. He was a real leader."

Both Caleb Choate and Godfrey Weatherby died in their early fifties, Mr. Choate at fifty-five and Mr. Weatherby at fifty-three. The same remark was applied to the lives of both men—"sad that it should be cut off ere it was fully rounded out." Godfrey Weatherby made this comment at the memorial service for Caleb Choate, and Frederick Choate, son of Caleb Choate, repeated it in his address at the memorial service for Godfrey Weatherby eighteen years later. A close friend, in commenting on his early death, said: "If Godfrey had lived he undoubtedly would have trained one of his sons to succeed him in company with William Pierce's son. But the children were so young when he died that his brother felt he would not have wanted his sons trained by the sort of managers that are in the shoe factories today. So he placed them in banks or bond houses where they would come under the influence of worthy, upright men, the type of men who used to be in the shoe business."

William Pierce started in the shoe business selling shoe laces for Caleb Choate and later became a shoe salesman. Following Mr. Choate's death, Mr. Pierce and Mr. Weatherby went into partnership and continued business under the firm name of Weatherby and Pierce for over thirty years, Mr. Weatherby responsible for the manufacturing end and Mr. Pierce for selling. Following the death of Mr. Weatherby, the firm liquidated.

Fred Jackson, of Jones and Jackson, in speaking about Mr. Pierce, of Weatherby and Pierce, said "what a good man he was. The firm liquidated because one partner (Weatherby) died and another had a bad heart. Mr. Pierce always felt responsible, for all his employees thought he was perfect, and he was worth half a million when he died."

A Greek shoe worker said: "I used to work for his father [Cabot Pierce] before he died. His father was a fine man. He was always a gentleman and would treat you right. He always paid more than anyone else."

The firm of Weatherby and Pierce shared this reputation, too. Among the items appearing in the *Herald* during the shoe strike and signed by the shoe workers, one observed: "The factory under William Pierce and Godfrey Weatherby's management was known the country over as a factory with ideal conditions between employer and employee. They (the employees) were always met more than half way and it was a privilege to work for them."

"The shoe business has changed considerably now," said a member of the upper-upper class. "Weatherby and Pierce was a fine concern. They made a very high-grade shoe, had the best workers in Yankee City and paid high wages. Both Godfrey and William were born and raised in Yankee City, and a lot of the shoe workers had been to school with them. Consequently they never treated the workers as employees but as friends."

This last statement was embodied in a story told over and over again during the strike: "Once Mr. Pierce met a shoe cutter named Sam Taylor on the stairs. Taylor said, 'Good morning, Mr. Pierce.' Mr. Pierce said, 'Good morning.' After he got back to his office he sent for Taylor and said, 'Sam, you went to school with me,' and Taylor said, 'Yes, Mr. Pierce.' 'Well,' said Mr. Pierce, 'you called me Mr. Pierce on the stairs just now. You always used to call me William, and I want you to continue to call me William just as you always did.' That was just a little thing, but subsequently whenever there was any dissatisfaction in the cutting room Sam would come down to the office and he and William would sit down at the table and settle the thing, each side giving in a little, and everyone would be satisfied. So there never was even the slightest hint of any labor trouble."

This parable was frequently told during the strike when all relationships between the owners and workers had been severed, the conclusion of the story serving as an eloquent moral lesson which the workers used to attack management.

"Every year they would shut down the plant," our interviewers were often told, in another story, "and the company would pay all the expenses of the employees for a day at the shore or in the country and everyone would have a share of chicken dinner costing two or three thousand dollars. The company had insurance for its employees, a benefit association, a hospital and trained nurse. . . . Of course all this cost a lot of money and added to the expense of making shoes."

In brief, the old owners were gods and not men. They had become heroes to labor as well as management. Where truth ends and idealization begins cannot be learned, but, fact or fiction, these memories stalked through the events of the strike like the ghost of Hamlet's father and motivated their sociological sons, their successors in the shoe business, to make decisions which were disastrous to them.

Before attempting to find out why the old owners became legendary heroes, let us look at the living managers who participated in the strike to discover further clues to this enigmatic situation where dead men often exerted more influence than did the powerful living.

2. Little Men and Aliens Run Things Now

THERE were an even dozen of them. John ("The Ram") Land, uterine nephew of the great Caleb Choate, leader of the shoe owners and partner of Cabot Pierce, son of the much beloved William Pierce, was the outstanding man on the side of the owners. His firm, Weatherby and Pierce, sold most of its shoes to Abraham Cohen and David Shulberg of the ABC Company. Cohen and Shulberg were "those New York Jews who are trying to run Yankee City." And then there were the Luntskis and Bronsteins, Jewish manufacturers from the suburbs of Boston.

Tim Jones and Fred Jackson were the local men of the firm of Jones and Jackson, but everyone knew that they took their orders from Abraham Cohen in New York. Finally, there was "Big Mike" Rafferty who must be thought of as more of a politician than an owner. He had been mayor of the town and was to be mayor again. These several managers divided into the "Yankee City men" and the "outsiders," and, when words grew hot, they were the "white men" and the "Kikes." As we said earlier, all of Yankee City was acutely aware of the powerful control the "outsiders" exerted over their lives, and everyone knew how little control the city had over the "outsiders."

"The men who run the shoe factories now don't even live in town," said a lady from an old family. "They drive down in the morning and spend the day making criticisms, then drive back at night, and no one knows how they live, how many automobiles or what kind of houses or how many children they have. They [the owners] don't know who their employees are or anything about them." Or, again: "At a time like this [during the shoe strike] psychology is very important. Obviously a workman has just as much right to live as a big fat manufacturer; but the manufacturers seem to think that their employees are way beneath them and don't require a living wage. This is a buyers' market, of course, and the manufacturer should have called his people together long ago and put his cards on the table. . . . But instead of anticipating any of these grievances the manufacturers went right on cutting and cutting and then they were surprised when the employees went out. Anyone could have known they were up for trouble."

"The Jews don't care a thing about people they employ or their families," said an upper-middle-class Irishman. "They think of people only as a means of making money."

"The manufacturers aren't real Americans," said a lower-middle-class woman. "They have the idea that the more they get out of us the better."

The several quotations above refer, of course, to the Jewish manufacturers who did not live in Yankee City and who had factories in other places. They were accused of "considering Yankee City only as a shoe town where they could obtain good quality of workmanship at a low labor cost." They were said to feel no responsibility or interest in the welfare of their workers. It was said the ABC Company was seeking a monopoly through price cutting of the turn-shoe trade and, therefore, "has every incentive to lower labor costs as well as disinclination through lack of personal interest to consider the welfare of the shoe workers—and they control 70 per cent of the shoe business."

When the "outside" manufacturers claimed to have a personal interest in their employees, the chairman of the Strikers' Emergency Committee, sarcastically reporting on a meeting where such claims were made, said: "The manufacturers explained to us workers what good fellows we were and how much they liked us shoemakers and how we had been one happy family for all these years and they just couldn't see why us shoe workers went out." When this story was told, as it often was, all the workers laughed knowingly and remarks of the following character were frequently made: "Yeah, we're all just one big happy family where all of us work for nothing," and "picture that goddamned New York Kike, he and his stooge, 'The Ram,' trying to get away with that kind of stuff. Why those bastards wouldn't give you the sweat off their balls!"

John ("The Ram") Land was the logical leader of the manufacturers' group, for he was the nephew of Caleb Choate. He had lived in Yankee City all his life and had always been associated with the shoe business there. His father and uncle started in the shoe business in 1866, Mr. Land's father putting in

$2,000, and Caleb Choate, $100. Mr. Land's father soon went into partnership with a local man while Caleb went on salary as a cutter with the expressed desire of learning the business "from the ground up." In the next year, Caleb Choate went into business for himself and forged ahead. Later Mr. Land was in business with his son, John, and within a few years before the strike John had entered into partnership with Cabot Pierce, son of William Pierce, who had invested a large sum of money in the firm in return for which his son was offered a partnership. So by tradition and experience Land should have been the leader of the manufacturers' group. And yet let us see what the various shoe workers had to say of him:

"Land is an awful hard man to work for," said one, "he doesn't care how little money he pays."

"Land has a hard reputation," said a Riverbrooker. "He won't pay. He doesn't like to pay his help but spends money on himself freely."

"They brought out a new-style shoe, paying eighty-nine cents a case," said an Armenian worker. "I worked six hours on the first case. Had to do it over four times so that I worked two days for eighty-nine cents and I quit. The boss [Land] came around and finally said to pay time instead of piece. You work and slave and can't do a thing. A man with a family can't quit; they say if you don't like it get out, there are plenty more. Land is the worst of them all, paying eighty-nine cents and others ninety-two cents to a $1.25 for the same work. Sometimes shoes look just alike but have different prices. The boss said, 'I can't help it, the buyer makes the prices.' We can't go on that way. I'd rather go out and starve."

"Land couldn't get in more workers," said a Greek worker, "because he has the reputation of only looking out for himself. People say no use working there."

An old and very much respected prominent Yankee City citizen (upper-middle) said: "He is a smart young man but the kind of person who whistles when he is on top and when he's beaten doesn't have much to say. He used to have meetings of his employees and I heard that he came in one afternoon to a meeting and said in a loud voice, 'If anyone wants to know who is boss around here just start something.'"

Shortly after the shoe strike, Mr. Land expressed his

opinions about the strike and the reaction of the public to it and to him. He said: "The strike was illegal anyway. My workers were all under contract with me—perfectly legal and binding and they had no right to walk out the way they did. I believe in organized labor; in fact, I signed an agreement with the shoemakers just two months before to work through a shop committee. I always tried to be fair to them; when times were good I paid them very good money, but what would you do in times like this?

"I was surprised that merchants and townspeople sided with strikers right from the start when they knew the odds the manufacturers were working under just to give people work. It would be cheaper for me to go out of business, but I feel I owed the working people something and I didn't like to fail them out of a job. I felt like getting someone to organize store clerks and see how the merchants would like that. The shoe would be on the other foot.

"I realize conditions are bad sometimes but usually they're unintentional and due to some foreman. The people up at the top just didn't learn the facts about it. I thought I had been helpful and fair. I don't think I'm God Almighty so I was perfectly willing to leave the decision to a third party. I didn't get any credit for it, I know. Everybody in town thinks I am awfully hard-boiled."

During the strike, Land defended himself in a letter to the strikers by likening his own behavior to that of Godfrey Weatherby. He said:

Godfrey Weatherby, highly regarded in this community and considered a fair man, taught me that the correct division or breakdown of the hundred-per-cent selling price was as follows: fifty per cent material, twenty-five per cent labor, ten per cent overhead, five per cent selling, five per cent discount, five per cent profit. I have checked up on this several times in all prices of shoes and with the National Shoe Manufacturers' Association and it is a fact that the breakdown of the sales dollar averages in these proportions, varying in some instances slightly where the selling is particularly low-cost because of connections, association, or some other condition permitting better materials to be used or higher paid labor.

This last condition has prevailed in Yankee City and because of a fortunate position with respect to marketing they have paid their labor in excess of thirty-five per cent of their selling price. I don't believe there is a single factory in Yankee City whose labor cost is not thirty per cent of their selling price.

Even the president of the union gave us evidence which belied Land's "confession" that maybe he was "hard-boiled." The union president said: "Land started out determined not to recognize the union. He said he had always dealt with his employees individually and intended to continue doing so. I said to him that I didn't blame him for his attitude but I'd like to have him think a little bit. I said I understood he had had trouble with his employees two or three times before and that he had won, but it must have cost somebody a lot of money. Maybe he'd win this time, too, but who would guarantee that next year he wouldn't have trouble again. Land said he supposed everyone in town was calling him The Ram, and I said, 'Well, you go along the street and people say, "Good morning, Mr. Land," but you know very well they are calling you something else behind your back. Maybe you enjoy that.' After a while I won Land over and he was very helpful.

"I doubt if we could have settled the strike without Land. I had heard that in Yankee City Land had the reputation of being a pretty tough customer but he certainly hasn't shown up in that light this time."

Thus, we have a picture of the reputation of the man who, by tradition and experience, seemed destined as leader of the shoe manufacturers' group. He was characterized as selfish and grasping, hard-boiled, and lacking in personal dignity. He was disliked and distrusted by a good many of his employees. His very nickname, "The Ram," suggested toughness and obstinacy. Contrary to his own words, he was believed to feel little or no responsibility for the welfare of his employees. His employees in town felt little or no loyalty toward him. And yet he must have had many qualities of leadership, judging by Nixon's and the labor arbitrators' comments concerning him and by the letters which appeared in the *Herald* over his signature during the strike.

As to Land's partner, Cabot Pierce, he seemed to take no

very active part in the management of the shoe business. "Cabot Pierce is all right but hasn't any say," said a shoe worker. "Cabot Pierce has no brains. He has been to about six schools, but he didn't learn anything. His father took him in, but he couldn't seem to amount to much. He used to take the men away from their work to play cards with them. When his father discovered it he scolded the men and didn't say anything to Cabot. . . . I think he won't last very long on this job."

And, from an interview with Pierce himself: "I feel badly about the small money people are making but what can we do? We have to make shoes just as cheaply as possible and there isn't any profit now in the shoe business."

During the shoe strike, Pierce appeared at the various conferences as a member of the manufacturers' group but did not participate actively in discussions; at least no mention is made of any contribution on his part. Cabot Pierce, resigned to defeat before management had lost the strike, sat on the sidelines frustrated and unhappy. The loss of thousands of dollars was on his mind, but the losses of the workers and his inability to think of anything he might do also pained him deeply.

Fred Jackson, of the firm of Jones and Jackson, on the other hand, over-participated in the strike with disastrous results to himself and interruptions to the negotiations in progress between the manufacturers and striking employees. Here is the story as Nixon, president of the union, told it to an interviewer (later verified from interviews with management):

"One of the manufacturers, Fred Jackson, a 'snappy' young fellow, came into a meeting and slapped a piece of paper down in front of me with a list of things Jones and Jackson proposed as an independent settlement. Jackson said, 'I'm going to make you eat that, Nixon.' And I said, 'Well, I don't happen to like paper, Mr. Jackson.' Jackson got very red and pulled a fifty dollar bill out of his pocket and slammed it down on the desk and said, 'You cover that, Nixon, and we'll go downstairs in the mayor's office and whoever comes out first wins.' I said, 'Don't be so childish, Mr. Jackson.' I only had about forty cents in my pocket at the time. The story got to New York and Jackson was called down the next day and fired."

According to Carter, the labor representative of the State Board of Arbitration, "the strike would have ended two weeks before if it hadn't been for Jackson. The manufacturers had all agreed to accept the union. Then someone started a rumor that the union intended to drive ABC out of Yankee City. Jackson heard it and, instead of investigating it or taking a trip to New York and discussing the matter, he got scared about his own job and wired ABC, who refused to sign up with the union. After ABC heard about Jackson's fuss with Nixon, Jackson got a long vacation. He was just a hot-headed young man with little experience or judgment."

The mayor said: "Jackson was a hot-headed young fellow with no experience of this sort. It was just crazy, and flourishing the fifty dollar bill was crazier, of course. Nixon handled him beautifully."

From another we learn that "the mayor followed Nixon almost to Boston to apologize for Jackson's action."

And a member of the Citizens' Committee observed: "Only a young man. He lost his head and tried to start a fight with Nixon. It shows pretty clearly he is not the type of man to win the confidence of his employees."

Jackson is the son of a small shoe retailer. He is said to have come from a Riverbrook family. His father set him up in business. He married a local girl who had formerly worked in a jewelry shop. According to one informant, "She flaunted her spending money, Pierce Arrow, chauffeur, and speedboat in a local beauty shop to the resentment of the other customers and staff who said that not so long ago she had been working for very little money herself." From behavior such as Mrs. Jackson's came the local opinion: "The manufacturers lowered wages but still get the same amount of profit, although they claim they are not making any money."

Of Timothy Jones, Jackson's partner in Jones and Jackson, we learn the following: "Jones and Jackson is a big plant. When they started, Jones had the experience and Jackson the money, or his father did. Jones lives across the river, and he has a lot of shoe workers for friends."

"Jones," said a shoe worker, "is a good man to work for. He came up from the bench himself, and he understands the shoe game. When they organized the workers and Jones encouraged them to walk out, he did it publicly. No secret about it."

A foreman of Land's said: "I know Tim Jones had been

stirring up trouble. I think it's pretty poor for a man in his position to try to agitate the workers even though he is in sympathy with them. And I think it is terrible the way he talked about Mr. Land. . . . Jones told his employees that the last [pay] cut was due to Land, when, as a matter of fact, Land hadn't cut at that time."

"ABC owns a controlling interest—fifty-two per cent—in Jones and Jackson," said Cabot Pierce. "Jones and Jackson are only salaried managers. They have no power."

Jones often went to Polock Lizzie's (a speakeasy frequented by workers) with his Riverbrook cronies. He had lived in Yankee City all his life and claimed to know everyone.

A significant extract from an open letter from the shoemakers in the *Herald* during the strike tells the same story about Tim: "There is one man in the shoe business who has been very fair. We are sorry to have to cause him any trouble. We are all for him and know that conditions are beyond his control."

This item seems to prove conclusively that, in the minds of the shoe workers, Jones was identified with them in spite of his position as part-owner and manager. In his own mind, Jones was a paid employee rather than an owner. He said that in spite of owning stock he was just a paid manager and had to take orders from New York without any say as to what should be done.

The president of the Moses Bronstein firm lived in Boston and had a factory in Boston as well as in Yankee City. The shoe workers said his attitude toward his workers could be judged by the story of the water cooler, reported by three different informants, apparently with some foundation. Bronstein installed a water-cooling system so that the employees could have cold water to drink in the summer. He bought it second-hand but ever since has been taking ten cents a week out of each employee's pay envelope to pay for it. The employees say they have paid for the cooler ten times over.

Bronstein's was called the "Penny Arcade" by the shoe workers because wage rates had been lowered so much.

A prominent Yankee City citizen said of Bronstein: "They say his sister comes into the factory and sells fruit and ice cream. Employees have to buy their food there, and she charges plenty high prices. Now you know Yankee City workers can't like that sort of thing . . . and the weekly ten cents out of each employee's pay envelope. That doesn't inspire a workman with confidence in his employer."

"At Bronstein's," said an Armenian worker, "the foreman doesn't like Yankee City men. They won't take them. They want somebody from Boston or Lynn. The union will prevent all that. We've been just like slaves."

The other Jewish concern in the city was Luntski's. One worker said, "Luntski is a pretty good Jew. An item in the *Herald* early in the strike said that the owners at Luntski's promised to give five hundred dollars to the Shoe Workers' Protective Union 'to carry on its work when an agreement is made.' Nixon reported that Luntski's wanted to sign up with the union without the 10 per cent increase. The strikers were willing, but Nixon said he had given his word that all would be treated alike."

"Big Mike" Rafferty (lower-middle) reported in his paper: "Luntski's say they will give the union a five-hundred dollar gift when it starts. Believe it or not. What big-hearted boys these birds are! When they first came to Yankee City they came in a Ford. Next they had a Packard, and now they are driving a big LaSalle. No wonder they can offer five hundred dollars as a gift. The union officials are hoping to be able to call their bluff."

Whatever the motive, Luntski's seem to have been perfectly sincere in their desire to sign up with the union. They were the first to sign the final agreement and the first, by several weeks, to reach an agreement with the union on wage-rate adjustments.

"Big Mike" Rafferty, president of the Rafferty Shoe Company, a very small enterprise, shows significant contradiction in his attitude toward the strike. A former mayor of Yankee City and local resident, Rafferty was the publisher of *Hard Facts*, a weekly filled with personal material about the reputable citizens of the city. People feared him because he was fearless and because his news items sometimes contained more than a grain of truth about matters which citizens might prefer not to have published.

During the strike "Big Mike" came out repeatedly for the strikers and against the manufacturers. Yet he complains in an issue of *Hard Facts* that the "Rafferty Shoe Company, of which 'Big Mike' himself is the president, treasurer and largest stockholder, was not invited to the Saturday conference which the local paper says was called by the Mayor. I think I should have been at least invited, but then you know our Mayor is such a gentleman. He wouldn't want a rough neck like me in his company.

"Perhaps the Mayor was afraid 'Big Mike' might let out some of the secrets. You know, folks, our Mayor was formerly in the shoe business and posed as a shoe manufacturer, but in reality he supplied about $80,000 in three years to let his brother-in-law play around with a shoe system."

At a later meeting of the manufacturers and strikers which "Big Mike" did attend, he seems to have comported himself with some dignity. According to an observer, he was quite reserved and said very little compared to what he usually said.

Carter, of the State Board, reported that at a meeting of the Board Rafferty sat alone on one side of the room. He sat quietly for half an hour, then got up and went over and pointed his finger at the other manufacturers who were sitting together and said, "Why, if I knew as much about shoe manufacturing as you fellows are supposed to I'd be ashamed not to pay employees more money."

Rafferty asked why "you shoe manufacturers who make your money in Yankee City don't sleep here, too?" He also charged that the ABC firm was trying to monopolize the shoe industry. He asked why "if the union could be recognized in Boston it couldn't be recognized in Yankee City?"

His last question made a strong impression on everyone. According to the mayor, "there was one thing the ABC people couldn't explain and that was the reason they finally agreed to the union, I think." He said, "They controlled the women's shoe part of the ABC shoe factory in Boston. They are under contract with a union there, affiliated with the American Federation of Labor, to be sure. But in Boston the ABC factory is under contract with this same union, the Shoe Workers' Protective Union, that the workers are striking for in Yankee City." The mayor said, "Cohen tried to get around it by talking about greater speed in Boston but that didn't explain it."

"Big Mike's" remarks in *Hard Facts* were most emphatic

regarding his stand on the strike situation. In an item headed "Labor Rebels!" he said: "The greatest labor demonstration in the history of Yankee City has been witnessed during the present week with the outpouring of workers from every shoe factory in rebellion against starvation wages and in many factories against unsatisfactory conditions. It is evident that the workers have the public with them and they should keep a united front so as to be able to demand a just and living wage as a reward for their labor. Pull together and *Hard Facts* will stand by you one hundred per cent. We don't sit on the fence."

Yet, after the strike was settled the Rafferty Shoe Company was the next to the last to sign the contract with the union, a week after most of the other factories had signed.

3. Structural Analysis of the Status of the Old and New Managers

SHORTLY after Caleb Choate's death, a number of prominent Yankee City men published a memorial volume which contained the usual words of high praise for a great man. Since these same words, unlike those of many memorial volumes, were said about him by ordinary men of the street, and, as we have said earlier, were used during the strike, it is important to examine them. Mr. Perkins Cantridge of Hill Street, a member of one of the oldest families of Yankee City, wrote: "Caleb Choate was one of the most remarkable men ever connected with Yankee City; a business man of liberal culture, of fine literary taste, gifted as an orator, in music and theatricals . . . He was an acquisition to any society. He honored any public station, however high. . . . He achieved more in his fifty years of life than most men can point to after marking a very old age. . . .

"He was identified with the public health of this city and was a conspicuous figure in all its great social functions as long as his health permitted it. He was a leading financier and a man who at once took and ever afterwards occupied a prominent position in this community. For years, by common consent, he was the leading man of the city. . . . Forcefulness of character made him the commanding spirit in every undertaking in which he shared and in every circle in which he moved."

Our analysis of Mr. Choate's participation in the community provides the crucial evidence on why Mr. Choate became the powerful symbol and collective representation which were used against the contemporary managers during the strike. We will briefly review some of the memberships that he had in the more powerful institutions of Yankee City.

In the business and financial sphere he was:

(1) owner and head of his million-dollar shoe company;

(2) president of one of the most powerful banks in the city;

(3) member of the Board of Trustees of the Financial Institute, a firm of the utmost prestige and power in the community;

(4) director of the Security Trust Company, another powerful financial institution;

(5) director of the Yankee City Gas and Electric Company.

He was involved in a large number of civic enterprises and was a member of many civic institutions:

(6) director and one of the founders of the city's most important hospital;

(7) director of the Public Library;

(8) member of the School Committee;

(9) trustee of the Revere Free School;

(10) president of the City Improvement Society.

He also took an important part in politics. He was:

(11) chairman of the Republican City Committee;

(12) member of the City Council;

(13) delegate to the National Republican Convention;

(14) mayor of the city.

Mr. Choate was also prominent in church and religious affairs. He was:

(15) president of the Yankee County Unitarian Club;

(16) president of the Yankee County Unitarian Conference.

He was a leader in fraternal affairs and was:

(17) Past Master of St. John's Lodge;

(18) member of several important fraternal orders.

Mr. Choate was an active member of some of the most exclusive clubs[2] of the city including:

---

2. The reader should consult pages 127–201 of Volume I of this series where he will find references to the power and social prestige of some of the associations to which Mr. Choate belonged.

(19) the Drama Club;
(20) the Thursday Night Club;
(21) the January Club;
(22) the February Club;
(23) the Lowell Club;
(24) the Country Club.

The evidence demonstrates that in all these organizations he was active and powerful. This brief survey of some of his participation in the community demonstrates that his activities ramified throughout the city and that much of the life of the city was centered in him. It also demonstrates that he accepted responsibility for the larger affairs of the community and helped integrate its activities, for he provided responsible leadership for the whole life of the community. "He was a man you could depend on."

Very much the same could be said about Mr. William Pierce and Mr. Godfrey Weatherby. They, too, were responsible elders of the city. Their factories provided jobs and wages. They were citizens of the town and men who felt obligated to it. Their membership in local institutions compares very favorably with that of Mr. Choate.

The essential point to remember about all three of these men is that they were subject to local control, because, first, they were dominated by local sentiments which motivated them "to take care of their own people"; second, they were under the powerful influence of the numerous organizations to which they belonged; and, third, their personal contacts with the local citizens directly related them to influences from every part of the city.

Mr. Cohen, Mr. Shulberg, Mr. Bronstein, and Mr. Luntski did not even live in the city. The workers knew or felt that the forces which controlled the local men did not control these "outsiders." The vast network of relations and memberships which made Choate, Weatherby, and Pierce local leaders, as well as local manufacturers, was reduced to a purely economic relation of employer and employee. It was that and nothing more. It is small wonder that the workers "gave the horse laugh when the managers talked about being good fellows."

Mr. Cohen and his group belonged to the last period in the economic evolution of Yankee City we spoke of earlier, that of

Big City capitalism, which had superseded the small town capitalism in the vertical structure of corporate enterprise and had extended on beyond Yankee City to the great metropolises. At the time of the strike the local men, although born and reared in Yankee City, were little more than the factory managers for Big City capitalists since they occupied inferior positions in this vastly extended vertical structure.[3] They were not in a position to take leadership; they were not in a position of great power where they were free to make the decisions which always characterized the lives of Choate, Weatherby, and Pierce.

Each of these local men felt what had happened very deeply, and some of them were explicit enough about it to say so. We knew some of them well. They were not weak men or unscrupulous persons as their opponents made them out to be. They had good personal reputations in the business world. Some of them had been trained by their own fathers to be community leaders, but their place in the new socio-economic structure of Yankee City prevented them from playing this role, and each in his own way contributed directly to the defeat of the managerial group. Part of their ineptness was due to their inability to measure up in their own minds to the great men of the past. This was a dead past, glorious and safe, when men knew themselves to be free men and Yankee City was "the hub of the universe." Clinging to the traditions of Choate, Weatherby, and Pierce, both workers and management longed to return to those days when it was possible for William Pierce, with all his power and prestige, to stop and gently chide Sam Taylor, the cutter, and he and Sam could talk about "the trouble in the cutting room." Power was under control and security was present then; manager and worker were part of a self-contained system in which each knew his part.

In these days of Big City capitalism, when Yankee City has lost control of its own destinies, few workers go up to the "Big Boss" to tell him about "what's wrong in the cutting room," and those who do are not considered respected friends at court of the

---

3. In 1945 all the local men who had been managers during the strike were no longer connected with the shoe industry in Yankee City, but one of them was employed in the shoe business elsewhere. Shoe manufacturing in Yankee City is still dominated by outsiders and ethnics.

workers but "stool pigeons who are getting theirs from management."

During the strike the local men cut poor figures as fighters for management's side. Tim Jones and "Big Mike" Rafferty openly lined up with the strikers. Local sentiment and the feeling against "the foreigners" were too much for them. They materially contributed to the workers' victory.

Jackson damaged the cause of management when he tried to fight the head of the union. Everyone said he blustered, and everyone said he acted badly when he challenged union leadership. Jackson was under the control of higher management and occupied an inferior managerial position where he had little freedom to assume command and take leadership. Yet he had learned from William Pierce when he worked for him how his kind of man should act, and he knew that an owner and manager should assume control. It seems a reasonable hypothesis that the conflict between his beliefs about how a man should act (how Mr. Pierce would do it) and what he was permitted to do by his status greatly contributed to causing his unfortunate act, an act which materially aided the union. He tried to take command in a situation where it was impossible, and he could only "bluster."

His antagonist, on the other hand, was "top manager" of the union. He did have power and he could make decisions. His beliefs about what should be done and his status were commensurate, and he used them to the greatest effect for the cause of the union.

To the workers, Mr. Land was everything that an owner should not be. His letters to the workers only embittered them. "His high and mighty attitude" was ridiculed because they believed he wasn't free and that he had to take orders even as an owner from his one big customer, Mr. Cohen. Cabot Pierce refused to take any action. He felt defeated before the strike began and acted accordingly, and thus gave no strength to the managers' side.

All of these local men knew somehow they were "not the men their fathers were," and the three dead men, symbolizing the glorious past, overawed and helped defeat them.

In the days before Big City capitalism took control, the local enterpriser was financed by Yankee City banks. These banks and other investment houses possessed more autonomy and prestige than they do now. In the development of Mr. Choate's shoe empire, local financiers played important and necessary roles and, at least part of the time, were silent partners in the business. Much of the wealth they derived from their investments was re-invested in Yankee City. The money was put into new enterprises, their own living, or in civic activities. Their white Georgian houses on Hill Street, whose gardens bordered those of the manufacturers, were majestic symbols of their power and prestige and forever reminded, and often reassured, everyone of the visible presence of these powerful and protecting men in Yankee City.

The Yankee City financiers, too, were men of responsibility, dominated by sentiments of local pride. They did well for themselves, but they also did well for the city. Perhaps the price was high, but the product bought by the rest of the community was substantial and of high quality. Their philanthropies, combined with their power and leadership, contributed enormously to the city's development and provided a firm foundation for the larger civic life of the community. Parks, libraries, hospitals, societies to help the unfortunate and aged, foundations to send young men to college, endowments of schools, churches, and many other worthy civic and public enterprises were granted and maintained by the money and leadership of the local financiers and manager-owners.

The ABC chain store with all its satellite factories, scattered through many cities and financed by several New York investment houses, is but one of many enterprises that these New York financial houses control. Their body of investment included Yankee City because it is one of the tens of thousands of living areas which make up the world. The flow of wealth from Yankee City's banks and factories, once a great local arterial system giving life and strength to the town, now has shrunk to an infinitesimal part of Big City, world-wide capitalism, where it has no vital significance.

The following account about the finances of the ABC Company, taken verbatim from a June 1945 issue of a large New York newspaper, supplies clear evidence for every statement which has been made here about the extension of the vertical hierarchy and the submergence of Yankee City into a very minor role in a world-wide financial-industrial structure:

A group headed by Oppenheimer and Co. and Brandeis and Son, and including the Stultz Co., has concluded an agreement for purchase of the majority of Lion Shoe Corp. stock, it was announced today.

Lion Shoe will be merged into its wholly-owned retail subsidiary, the A.B.C. Shoe Corp., with subsequent public issue of securities of the latter company.

Abraham Cohen, associated with the companies in an executive capacity for more than 20 years, will be elected president and general manager. Frederick Stultz, president of the Stultz Co., will be made chairman of the board.

The A.B.C. Shoe Corp. owns a number of factories equipped to manufacture 20,000 pairs of shoes daily and operates a chain of 110 stores in 56 cities.

Decisions on these high levels of national and international finances, which vitally influence Yankee City and its chances of survival, can be, and are being, made, which totally disregard all Yankee City's needs and vital interests. It is certain that decisions charged with ruin or success for the economy of Yankee City and the stability of the lives of its people are made by men at the policy level of such international financial houses who do not so much as know the name of Yankee City and who, beyond all doubt, do not care what happens to the town or its people.

The men of yesterday are dead; but their "souls go marching on" in the memories of the living, and Mr. Choate, Mr. Weatherby, and Mr. Pierce are collective symbols of that lost age when the prestige and power of local financiers and local producers "took care of our own people." Admittedly, these men did it for a high price, but at least the workers and ordinary town people were more highly rewarded by Mr. Choate than by the banking houses of New York and London. Today even the name of Yankee City is not known to those whose financial power often controls decisions of the utmost importance for the town. It is not difficult to understand why the symbols of Mr. Choate, Mr. Weatherby, and Mr. Pierce collectively represented small city finance and its lost rewards and satisfactions, as well as the one-time security of local ownership of the factories. Given the significance of the symbols, it is obvious why they became powerful allies of the strikers and helped force management into submission.

From this analysis, several important propositions can be offered which contribute to our understanding of why the strike happened and why it took the course it did. The vertical extension of the corporate structure of the shoe manufacturing enterprises had pushed the top of the hierarchy into the great metropolises and, in so doing, had brought in "outsiders" who were "foreigners" in culture and lacking in understanding, feeling, and prestige for the local workers and for the town itself.

This extension of the industrial hierarchy reduced the local men to inferior positions in the hierarchy where they were incapable of making decisions and could not initiate actions which would give them the power of leadership for the workers and for the rest of the town. Reducing the local managers to inferior statuses in the factory contributed to their lower social-class ranking in the community and thereby greatly reduced their strength as leaders and men who could form community opinion in times of crisis when the position of management was threatened. They could no longer lead the workers or the community. Because of the inferior position of the managers, those men in the community who would have once been their natural allies and who enjoyed top social-class position were now above them and shared none of their interests, were hostile to them and friendly to the workers. The vertical extension of the corporate structure of the shoe business introduced owners into the community who had only economic memberships, whereas in the previous period of local control an owner had power and leadership in all of the important institutions.

The longing for the idealized past when men had self-respect and security was symbolized in the three dead owners; and these symbols materially aided the workers in defeating management since the workers and management felt that the present men could not match the "gods" of the past. The workers and managers in the shoe industry had lost their sense of worth and mutual loyalty. No longer were they men who had a common way of life in which each did what he had to do and, in so doing, worked for himself and for the well-being of all.

IX

THE WORKERS LOSE STATUS IN THE COMMUNITY

1. *A Coming Industrial Proletariat?*

THE last part of our analysis of the strike tells the story which, if generally true—and it seems to be, is a cause for deep concern to those who believe in American democracy. We ask ourselves the questions: How do the shoe workers compare with the other people of Yankee City when you examine what they are like in the community? Are they different or largely the same? At the outset we were of the opinion that they were not different from everyone else, but an analysis of their participation and that of other people of similar status in the community showed that our assumption was false.

There is clear evidence to support the hypothesis that workers are losing status as a group. We found this out by first examining over seventy thousand memberships in the various groups composing the institutional life of Yankee City.[1] After looking at the memberships of all the people in the community, we separated the membership participation of the workers and found that it differed very substantially from the interaction of the total group in the social structure. If this is true (and we believe we can demonstrate that it is) there is one more factor accounting for the general dissatisfaction among the workers which contributed to the explosion when the strike began.

Our statistics give much evidence to show that the social life of the shoe workers outside the factory is noticeably different from the life of total adults of comparable status in the community. With a large and probably increasing percentage of those who earn their living engaged as workers in great American industrial enterprises with a developing social life that is increasingly different from that of other adults of comparable status, it may indicate that our industrial system is molding

---

1. See *The Status System of a Modern Community,* "Yankee City Series," II, 1–71.

a new category of class relations. If those members of the three lower classes who work with the machines of industry act and feel alike in their social relations, we may be forced to recognize that a new social pattern of behavior is being created which will cut across, or considerably modify, the structure of our social classes.

This differentiated social behavior of shoe operatives does not seem to bear any relation to their economic status; it seems to be related to the fact that they are all engaged in a type of work in which they are thrown into a far closer relational pattern than are others of similar social status in the total community.

This emerging industrial working class or working group appears to reflect, from the behavior of these shoe operatives in the life of the community, some common need which may be supplied only by a pattern of behavior which is common to the group and has a group characteristic. It may be that, by the blockage that prevents them from rising in the industrial status hierarchy, they find a security in cementing the solidarity of the group itself.

If our evidence tends to confirm an increasing solidarity of industrial operatives as a group, such solidarity implies that there must be some threat to those who comprise the group. Our data indicate that the shoe operatives have been blocked in their opportunity to obtain more lucrative jobs and jobs of higher status in the factory.

The emergence of an industrial working class seems, therefore, to be derived from the increase of the subordination of industrial workers by the division of labor and the demands of business efficiency with its consequent blockage of upward mobility in the factory system. On the other hand, the increase of the solidarity of this industrial working class or working group suggests that its members believe that only by some form of collective effort can they combat their subordination and be given a chance to get ahead.

Our interest is in the effect on community life of the factories as *social* institutions and what has happened to the workers as citizens. The fundamental question we now ask is: "What effect, if any, does working in a shoe factory have on the participation of operatives in the general social life of the

The Workers Lose Status in the Community

local community?" A large portion of the three lower social classes in Yankee City was made up of the shoe operatives. If we can demonstrate that the social behavior of these operatives differed significantly from that of the general population similar to them in social characteristics, we shall be justified in attributing these differences to the effect of the shoe factories on the operatives' lives in the larger social context of community activities. If this effect appears to be considerable, then the shoe factories are a powerful factor in molding the social structure of Yankee City in its broader—not merely economic —aspects.

To solve our problem, we have compared the social behavior of shoe operatives with that of the total Yankee City adult population[2] of the three lower classes. This comparison has been made in terms of five distinguishable types of social behavior which the larger research has shown to be basic determinants of the social structure of the community. They are: (1) associational memberships, (2) clique memberships, (3) family structure, (4) religious participation, and (5) political participation. We shall state exactly what is included in each of these types of social behavior.[3]

Before the reader can understand the analyses and comparisons in this section, he must fully comprehend several concepts we use throughout. The most fundamental of these is the positional system.[4] Chart IX represents the primary concepts involved in analyzing social structure in terms of social positions or statuses. The six horizontal levels represent the six social classes or strata into which Yankee City society may be divided. They are upper-upper (UU), lower-upper (LU), upper-middle (UM), lower-middle (LM), upper-lower (UL), and lower-lower (LL). Every individual in the community has a place in one or another of these social strata (except for a

---

2. By an adult we mean any person eighteen years of age or older. In the three lower classes in Yankee City, individuals assume adult status at about eighteen. This also happens to be the age of the youngest shoe operative.

3. In the larger study, school participation was also examined. This was omitted here because none of the shoe workers attends school.

4. The positional system for the community as a whole is fully treated in Volume II of this series. The concepts involved are not abstruse, but their formulation is new in sociological literature. They are basic to the understanding, not only of the shoe worker's place in the social structure of Yankee City, but of the social structure of the whole community.

comparatively few who are in the process of passing from one social class to another, either up or down); therefore, the horizontal position of every individual in the society may be schematically represented. But every individual has various kinds of social relations with other individuals both in his own stratum and in other strata. In every combination of such relations, the individual has a different social position. Every type of intra- and inter-stratum social relation we observed in our study of Yankee City is represented in Chart IX by the small double-pointed arrows connecting pairs of small circles. The circles themselves represent social positions, which are the structural aspect of social relations. The circles ("positions") are numbered for easy reference. If Mr. James Smith, a lower-middle shoe worker, for instance, is in a social situation where he has direct social relations with an upper-upper person, but with no one in any other social stratum, we say that in this situation the lower-middle man occupies Position 48, while the upper-upper person occupies Position 8. Again, for convenience of reference, we have numbered the vertical columns of the positional diagram, Chart IX; in the present example, the relationship between the lower-middle man and the upper-upper person would be referred to as a relationship of Class Type 8. The thirty-four different class types of relationships which were identified in our analysis of Yankee City social structure are thus indicated in Chart IX. They show the extreme range of social contact for each position.

Almost every individual in the society has a great many social relations with various individuals and groups belonging to his own and other social strata. Each of these sets of relationships can be labeled by stating its class-type, or range of contact; and the social position of any given individual in that set of relations can be identified by stating his position number as determined by the coördinates of the individual's social stratification and the class-type of relationship involved. Thus, Mr. James Smith belongs to the American Legion, which is an organization of Class Type 11, since it draws its members from all social classes or strata. As a Legionnaire, Mr. Smith occupies Position 51. He might, in fact, if his range of social participation were wide enough, occupy all the lower-middle positions from 45 to 64 inclusive. It would be very unusual,

CHART IX

The Positional System of Yankee City

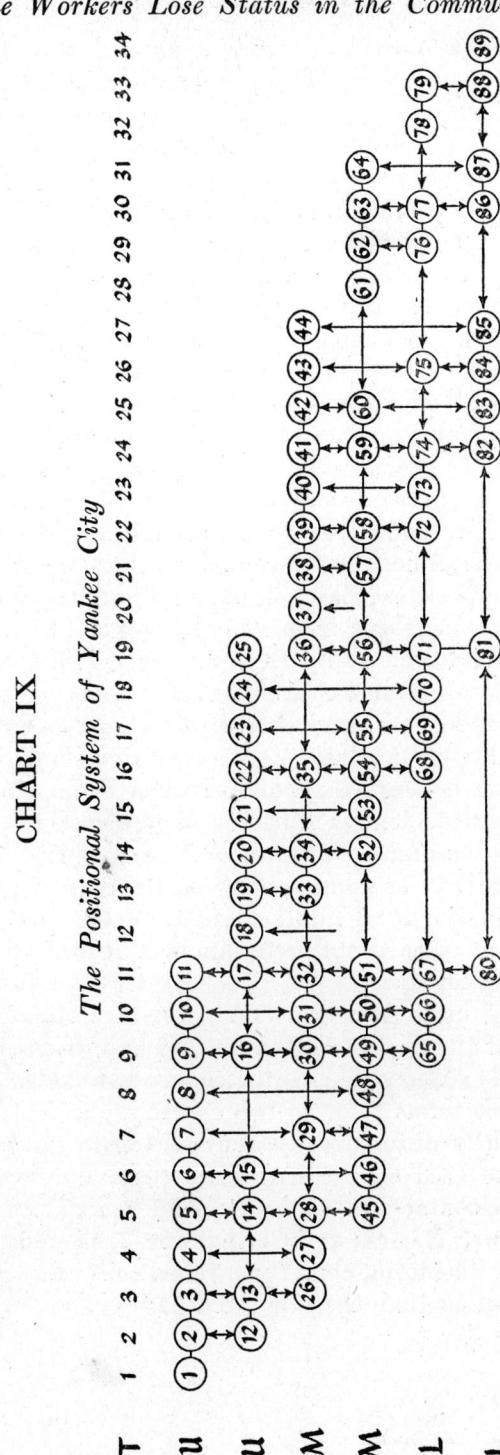

however, for one individual to participate in so many different kinds of social relations. The different social personalities of different individuals in the same social stratum can be indicated, in fact, by showing the difference between their individual positional characteristics.

A point which must be made clear before the reader can understand the material we present in this section is that *one* individual (such as Mr. Smith) may, and in almost all cases will, occupy several different social positions, i.e., be represented in several positions in a diagram like Chart IX, depending on the various kinds of social situations in which he participates. Each of these various positions we call a *membership*. An individual will also hold two or more memberships in a single position if he has social relations with two or more sets of individuals who cover a similar range in social stratification. Thus, since Mr. Smith is a "regular" Democrat as well as a member of the American Legion, he holds two memberships in Position 51, because the political party and the Legion both draw their memberships from all social strata. Some churches also draw their membership from all social classes; since Mr. Smith belongs to such a church he has a third membership in Position 51. It is clear, therefore, that in most extensive examinations of the social relations of a number of individuals, the number of *memberships* they hold will exceed the number of individuals studied. Unless otherwise specified, our numerical references are to number of memberships of various sorts held by shoe operatives as compared with the number of memberships held by the total adult population of the three lower classes. References will not be to numbers of *individuals* unless we specifically so state.

One more point must be clarified before the reader can understand some of the important tabulations and discussions which appear in this section. This is the concept of *extremes* of class contact. These terms refer to two extreme positions in a class type—a slightly different way of referring to the same social phenomena covered by the term *class type*. For example, the extreme class contacts of Class Type 11 are the upper-upper and lower-lower classes; for Class Type 5, upper-upper and lower-middle. The term, class type, refers to *all classes* involved in the type rather than the two extremes.

A shoe worker, Jim Smith, was a member of ten institutions. Each of these memberships belonged to a particular class type and position. As a lower-middle-class member of the American Legion and the Democratic Party, he had contact with the two extremes of the class system, men from the upper-upper and lower-lower classes. In addition to the Legion and his political party, Mr. Smith belonged to seven other institutions. The extreme class contacts of four of these were upper-upper (UU in Table 4), two went to lower-upper, one to upper-middle, one included only lower-middle, two extended down to the upper-lower, and five to lower-lower (see Table 4 for all of these).

TABLE 4

Sample Tabulation of Extreme Range of Social Participation

	UU	LU	UM	LM	UL	LL	Up	Same	Down
LM (10) No.	4 40.0	2 20.0	1 10.0	1 10.0	2 20.0	5 50.0	7 70.0	1 10.0	7 70.0

We are now in a position to measure how much contact Jim Smith had with members of all classes of Yankee City society. We can find out if his social position afforded companionship and participation only with those below him or if his membership in these ten organizations pulled him to levels higher than his own class position.

When a membership gives Smith class contacts on both sides of his own class, such a membership obviously gives him *two* contacts, which we have counted as *two*. But when a membership extends in only *one* direction there is but *one* class contact beyond his own class, and that has been counted as *one*.

Jim's memberships were of the following class types: four extended in only *one* direction (thus, a total of four contacts); five extended *two* ways (therefore, a total of ten contacts); and one (see Position 61 in Chart IX) went neither way (a total of one). This makes a total of fifteen contacts; seven were above (see right hand column of Table 4), seven were below, and one was neither.

If we divide Smith's ten memberships into the extreme contacts with each class we acquire an index of the extreme range of his social contacts and have a way of comparing his social range with that of everyone else in his class and in all other classes in the society.

The most important comparisons we make in this section between the social behavior of shoe operatives and that of the total adult population of the three lower classes hinge on analyses like that for Jim Smith. The explanation we have just given is the interpretative key to what follows. This type of analysis can be applied to social structures of every type however formal or informal they may be. We shall apply it first to participation in associations, then to clique, family, religious, and political participation, in that order.

We are now ready to measure the loss of social status by the shoe workers.

By the term "association" we mean a more or less formally organized group, such as a club, as distinct from the completely informal groups we call cliques. Membership in an association is voluntary; it therefore reflects to a high degree the individual choice of the person joining. If a membership were mandatory for any individual, its significance for our purpose would not be as great.

Significant differences are noticeable between the kinds of associations in which the shoe operatives participated and those to which the total adult population of the three lower classes belonged. Table 5 indicates certain material variations in the percentage of class types of memberships held by individuals of corresponding social class in the two groups. Where the variation in percentage is great, the percentages are italicized for emphasis. The significance of the variation between the associational behavior of shoe operatives and that of all adults in the three lower classes of the community is best analyzed by examining independently the memberships held by members of the lower-middle, upper-lower, and lower-lower classes in each group before making any generalization as to the meaning of the total variation.

The shoe operatives of the lower-middle class held a smaller percentage of memberships in associations which had members rated above the lower-middle class and a larger percentage in associations that were rated downward from the lower-middle class (64.3 per cent and 77.7 per cent respectively) than did the individuals of the same social stratification in the adult group with which we are comparing them (91.1 per cent and 63.4 per cent). The membership in associations which include members rated as high as upper-upper was markedly less for the shoe operatives than it was for the whole adult group (14.0 per cent against 35.6 per cent).

TABLE 5

Associational Membership—Extreme Range

		UU	LU	UM	LM	UL	LL	+	0	−
Shoe operatives										
(157)	LM	22	39	40	0	7	115	101	0	122
	%	14.0	24.8	25.5	—	4.5	73.2	64.3	—	77.7
(404)	UL	28	46	115	214	0	389	403	0	389
	%	6.9	11.4	28.5	53.0	—	96.3	99.8	—	96.3
(293)	LL	15	9	39	225	5	0	293	0	0
	%	5.1	3.1	13.3	76.8	1.7	—	100.0	—	—
Adult Yankee City										
(3,196)	LM	1137	773	1001	29	416	1610	2911	29	2026
	%	35.6	24.2	31.3	0.9	13.0	50.4	91.1	0.9	63.4
(2,314)	UL	332	489	1014	459	5	2129	2294	5	2129
	%	14.3	21.1	43.8	19.8	0.2	92.0	99.0	0.2	92.0
(825)	LL	137	74	255	291	60	8	817	8	0
	%	16.6	9.0	30.8	35.3	7.3	1.0	99.0	1.0	—

Among the upper-lower class of shoe operatives the percentage of memberships in associations having members below and above the upper-lower class was not materially different from the comparable percentage for the total adult group. A significant variation is noticeable, however, when we consider that the class types of associations whose members extended to the upper-upper, lower-upper, and upper-middle, respectively, were more numerous for the general adult group than for the shoe operatives. But the class type of association which had no members above the lower-middle was over two-and-one-half times as great for the upper-lower group of shoe operatives as for the total adult group (53.0 per cent against 19.8 per cent).

The variation between the association memberships of the lower-lower group of shoe operatives and those of the corresponding group of adult members of the society is similar to upper-lower. Over three-fourths of the memberships of the lower-lower group of shoe operatives were in associations which did not have members above the lower-middle class, as against about 35.0 per cent of such memberships in the total adult group. In the types of associations which have members rated as high as upper-upper, lower-upper, and upper-middle, the percentage of total memberships was again much higher for the total lower-lower adult group of the total society than it was for the equivalent group of shoe operatives.

The foregoing analysis of the data, summarized in Table 5, shows conclusively that the shoe operatives, much more than the total adult individuals of the lower three classes, tended to restrict their associational memberships to those associations which were oriented toward the lower classes. A decidedly smaller proportion of shoe operatives belonged to associations which included upper-upper or lower-upper members. Particularly striking, too, was the concentration of associational memberships among upper-lower and lower-lower shoe workers in associations whose topmost members were lower-middle. This suggests that as far as associational behavior goes, there was a tendency among shoe operatives toward the coalescence of the three lower classes to the exclusion of all classes above lower-middle. The figures for the total adult individuals of the three lower classes suggest no such coalescence. There was evident in this group a tendency among upper-lower individuals to concentrate in associations whose members rated two strata higher, in upper-middle; and for lower-lower individuals to concentrate in associations having either upper-middle- or lower-middle-class members.[5]

The clique, which we shall consider next, is (as we are using the term) a group of individuals, usually few in number, who are not formally organized into an association, and have no officers, charter, or specified purpose for being. The members

---

5. The union memberships are not included in the figures for the total community or for the shoe workers.

of a clique have close personal relations with one another, and as a result they often manifest greater solidarity than do members of associations.

The following table, prepared in the same manner as the table covering associational memberships, gives the extreme range in social stratification of individuals in the different types of cliques, and compares shoe operatives with the total adult lower-middle, upper-lower, and lower-lower population.

TABLE 6

Clique Membership—Extreme Range

		UU	LU	UM	LM	UL	LL	+	0	—
Shoe operatives										
(156)	LM	1	10	44	11	91	42	55	11	133
	%	0.6	6.4	28.2	7.0	58.3	26.9	35.2	7.0	85.2
(528)	UL	1	16	104	228	111	244	349	111	244
	%	0.2	3.0	19.7	43.2	21.0	46.2	66.1	21.0	46.2
(297)	LL	0	0	16	135	82	64	233	64	0
	%	—	—	5.4	45.5	27.6	21.5	78.5	21.5	—
Adult Yankee City										
(4,219)	LM	69	124	1585	608	2018	918	1778	608	2936
	%	1.6	2.9	37.6	14.4	47.8	21.8	42.1	14.4	69.6
(3,834)	UL	38	152	819	1855	549	1455	2864	549	1455
	%	1.0	4.0	21.4	48.4	14.3	37.9	74.8	14.3	37.9
(1,415)	LL	0	0	109	575	383	348	1067	348	0
	%	—	—	7.7	40.6	27.1	24.6	75.4	24.6	—

Since a clique is, by our definition, an intimate group of individuals, its membership is more frequently confined to a single class or adjacent classes in the social strata than is associational membership. This is brought out in Table 6, in which we have again italicized certain particularly significant percentage figures. In the clique memberships, there are evident differences between the extreme range in social stratification of the cliques to which shoe operatives belong and the range of memberships of the total adult population of the three lower classes.

This table shows that in the lower-middle class the percentage of memberships in types of cliques which have members above the lower-middle class was less among the shoe operatives than among the total adult group, and greater in types of cliques having members below the lower-middle class. This holds true also with reference to upper-lower-class participation. Among the lower-lower class a slightly higher proportion of shoe-operative memberships were in types of cliques having members above the lower-lower class than among the total adult group.

In general, these characteristics of shoe-operative memberships in cliques compared with the clique memberships of the total adults of the three lower classes agree with our findings regarding the comparison of association memberships: shoe operatives tended to orient their social relations toward the lower levels of the social stratification to a greater extent than did comparable segments of the total adult population. There is also observable in clique behavior the tendency of the shoe operatives to coalesce in the three lower social classes to a greater extent than did the total adult population of these classes: the lower-middle shoe operatives oriented downward, and the lower-lower shoe operatives oriented upward to lower-middle but not above, to a greater extent than is true of the total population in these classes. The general orientation of all shoe operatives was toward upper-lower, a phenomenon not to be observed among the total adult population of the three lower classes.

Family membership does not differentiate shoe operatives from the total adult population of the three lower classes as clearly as do the corresponding characteristics of association and clique memberships. What tendencies we have been able to discover, however, have indicated that, on the whole, comparisons of family memberships agree with the more clear-cut findings regarding associations and cliques: there was a tendency for shoe operatives to orient themselves toward the lower end of the social scale, as opposed to the tendency of the total adult population of the three lower classes to strive for upward orientation.[6]

The significant results of our comparison of religious faith between shoe workers and all members of the three lower classes

---

6. The full details on the family are given in Appendix 4.

of the society are: (1) a considerably smaller proportion of shoe operatives professed religious faith; (2) although Catholic and Protestant members were almost exactly equal for the total group, there were half again as many Catholics as Protestants among the shoe workers; (3) the proportion of Greek Orthodox among the shoe workers was much greater than among the total population of the three lower classes.[7]

We find that lower-middle shoe operatives participated to a greater extent than did the general lower-middle population in social structures which extended below lower-middle in social stratification; that a larger proportion of upper-lower shoe operatives than of the upper-lower general population had memberships in social structures which extended downward to lower-lower; and that memberships of lower-lower shoe operatives were definitely found more frequently in social structures with members of higher social stratification—up to and including lower-middle—than were memberships of the general lower-lower population. On the whole, this last consideration indicates that lower-lower shoe operatives tended less than the total lower-lower population to regard themselves as members of a circumscribed lower-lower stratum; these shoe operatives tended to range, in their social participation, through all three of the lower classes, but not above. Detailed comparisons of upper-lower memberships reveal parallels to the lower-lower comparison. Similarly, lower-middle shoe operatives tended to relate themselves downward in the social scale much oftener than did the general run of lower-middle Yankee City citizens.

The workers of Yankee City were able to strike, maintain their solidarity, and in a sense flee to the protection of the unions because the disappearance of craftsmanship and the decreasing opportunities for social mobility had made them more alike, with common problems and common hostilities against management. The craft differences had been wiped out, and occupation mobility in the craft hierarchy and, secondarily, social mobility in the community had been stopped. The workers felt even more alike and were increasingly motivated to act together because their new occupational status had contributed to their downward orientation in the community.

7. For full details about the church see Appendix 3. Political participation is given in Appendix 3.

There is no doubt that each worker's uneasiness and reasoning about what was happening to the status of his family and himself—a situation which he meagerly comprehended and which was almost beyond his ability to communicate coherently to his fellows—whipped him into attacking the owners, who provided visible targets and could be held responsible for the loss and degradation of the worker's cherished way of life.

Americans, and American workmen particularly, find it hard to think and talk about status and social class. Our reigning ideology of "I'm just as good as anybody else" denies the existence of rank; "what ought to be" is substituted for "what is" in the minds of the people. Class and similar status hierarchies are denied. For those in the lower half of the status system the assertion that the democratic dogma is true and that everyone is equal is deeply satisfying and compensates them for the indignities suffered in their subordinate position. Yet each feels the pressure and importance of social status so keenly that the slightest drop in rank which a man or his family suffers always causes a strong and sometimes violent reaction.

The downward orientation of the workers supplied part of the necessary pressure on them to cause the strike. The discontent and resentment accumulated from the experiences of a generation of downwardly oriented workers, we believe, greatly contributed to its outbreak. The strike gave shoe workers an opportunity to express their long pent-up feelings of frustration and loss of self-respect.

2. The Causes of the Yankee City Strike

BEFORE drawing our final conclusions on why men strike and why they join unions, we must focus our attention on what we have learned about the causes of the strike in Yankee City. We know that the factors were multiple, as they are in all complex human situations, and that there were a number of basic factors sufficiently powerful to explain why the strike occurred and unionization took place.

The economic basis of the conflict between the two groups appears to lie in the divergence of the systems of logic under which each carries on its work. The logic which dominates the working relations of the supervisory group may be called the business (or profit-making) logic. Under this logic the one controlling aim of those in charge of the shoe factory is to produce shoes at the lowest possible cost and highest profit. It may be said, in general, that all the members of the supervisory group subscribe to this logic and feel their personal interests to lie with the management and owners of the factory in the pursuance of its dictates.[8] This behavior is to be expected on the part of everyone in the supervisory hierarchy because the business status and hope of advancement of each individual depend upon the contribution he makes to efficient production. Individuals in this group compete with one another for advancement in the hierarchy; the individual chosen for promotion from among a number of persons of approximately equivalent status in the hierarchy is usually the one who, in the opinion of his superiors, contributes more than his fellows to the efficiency of production. All the members of the supervisory hierarchy, therefore, may be said to be primarily motivated in their working relations by a common desire to maintain or increase the profits of the factory.

The shoe operatives, on the other hand, are not interested in either maintaining or increasing the profits returned to the factory owners; their principal economic concern is that the factory should remain solvent and continue to afford them jobs. It is inevitable, therefore, that desires which motivate the working behavior of operatives conflict, at times, with the profit-making motive which largely governs the working behavior of the supervisory group.

In this conflict of interests, the operatives, though they far outnumber the supervisory group, occupy, as we have said, a subordinate position. Their subordination has resulted largely from the extensive division of labor which has occurred in the technological end of shoe manufacture. The division of labor serves the management group, not only by increasing the rate of production, but also by simplifying the problem of controlling the operatives. In the days when shoes were made by hand and the master craftsman performed or supervised the operations by which raw or semi-finished materials were converted

---

8. When interviewed as individual citizens almost every person holding a supervisory job in Yankee City shoe factories was interested in, and concerned for, the welfare of the shoe operatives as human beings. But in his character as a member of a managerial hierarchy each had to subordinate his personal feelings to the requirements of the business logic.

into finished shoes, the technological expert was in a strong bargaining position because there was a ready market for his abilities and he was protected against technological competition by the years of training required to gain expertness in the craft. He had a great deal of freedom of action in his technological activities and also in his working relations with the supervisory group.

In the modern, largely mechanized shoe factory, jobs have been extensively simplified, routinized, and standardized. In the process of manufacture, a pair of shoes, instead of remaining largely in the hands of one or a very few experts, now passes through the hands of a large number of operatives, each of whom performs only a small part of the manufacturing process. In addition, many of the jobs consist primarily in feeding materials to machines which perform the actual processes. The division of labor in modern shoe factories has thus reduced tremendously the freedom of action of operatives in the performance of their jobs. Each job is so simplified that the technological behavior of a given operative is strictly limited in range; in the case of mechanized jobs, the operative must even accommodate his permitted actions to the tempo and rhythm of the machine he operates. The exceedingly limited scope of the technological activity permitted the modern shoe operative makes it relatively easy for a foreman to maintain discipline because he can immediately detect any deviation from the permitted patterns of behavior. The character of the relations between a foreman and the operatives under him is likely to be much different today from what it was in the handicraft days. Today a foreman need seldom be a working boss, familiar with the techniques of shoemaking; he may be merely a disciplinarian, may even have been hired from some other industry because of his ability to control subordinates.

The routinization of jobs also simplifies control of workers in another way. The individual operative today does not have the feeling of security that the oldtime craftsman derived from his special technical abilities. In most cases, today's operative is aware that only a comparatively brief training period protects him in his job from a large number of untrained individuals. The members of the supervisory hierarchy are also well aware of this fact. The psychological effect of this result of the division of labor is to intensify the subordinate position of the individual operative and to make him submit the more readily to the limitations on his behavior required by the supervisory group. Furthermore, some foremen consciously hire individuals toward whom they feel superior in every way and discriminate against persons they call "educated." They want operatives who will do as they are told without asking questions. Creative thinking on the part of operatives is discouraged by the supervisory group.

Members of the latter group, on the contrary, are expected to think creatively about their jobs, within the framework of the profit-making logic, and are rewarded for it. The type of individual who is sought after and encouraged in the supervisory group is discriminated against and discouraged in the subordinate, operative group. Add to this the fact that the working behavior of operatives is strictly limited to the performance of well defined techniques, and the further fact that foremen need only know how to control other individuals, and it will be readily understood that it is almost impossible for an operative to get out of the subordinate into the superordinate group. He has the opportunity neither to prepare himself for a foreman's job nor to demonstrate his ability to perform such a job.

The great and powerful motivations provided by social mobility in the class structure are largely denied the operative since his low-placed jobs are no longer in a craft hierarchy. The worker's wife and family are usually condemned to staying at a lower social level, knowing that they have little opportunity to rise and to get their share of the glory of the American Dream.

There is a second factor, closely related to the conflict of interests between the supervisory and operative groups, which is important in the social organization of the shoe factories of Yankee City. This is the break in the skill hierarchy. In the days when shoes were made by hand there was a well defined technological hierarchy. There was plenty of opportunity for advancement through individual effort in the technological end of the shoemaking business.

But all this is changed. There are no high-skilled jobs in the Yankee City shoe factories[9] (except possibly for an occasional designer). The high-skilled jobs associated with shoe manufacturing are now found in the research departments of allied but separate industries. The men who hold these high-skilled jobs are trained in technical schools; they are not drawn from the shoe industry itself. Shoe operatives have even less chance of getting such jobs than they have of getting into the supervisory group of the shoe factories, for they have no opportunity at all of getting the specialized technical training for these jobs in the shoe factory.

At the same time, the designers and manufacturers of shoe-making machinery have progressed so far in the analysis and consequent mechanization of shoe-manufacturing processes that most of the technological jobs in the shoe factory itself have been reduced to a common low level of skill. The breakdown of jobs in the shoe factory has been so thorough, in fact, that almost never does the learning of one job prepare an operative for another. There is, in other words, almost no opportunity for a shoe operative in a Yankee City factory to increase, through his individual efforts, either his security or his prestige, in the technological end of shoe manufacturing.

Yankee City shoe operatives, then, are unable, by individual effort in working relations, to increase either their security or their prestige. What is denied them as individuals in working relations, they have sought to gain by collective action through union membership.

It is significant that the union joined by the Yankee City operatives at the time of the 1933 strike was of the so-called industrial type rather than a craft union. Craft unions flourished in the shoe industry in the days when there were effective technological hierarchies in the industry. They still flourish in industries which have hierarchies of technological jobs. In the shoe industry, however, and other industries where the technological hierarchies have broken down, the modern industrial union, in which every technological worker in the industry is eligible for membership on a basis of equality with every other individual, is functionally consistent with the working relations

9. Designing is still an occasional exception, but even this job is losing its importance in the shoe factories themselves.

that actually obtain among operatives. When the technological hierarchy has been destroyed through division of labor and mechanization, all workers are actually on a common footing and have common interests in opposition to the management group. From the sociological point of view, therefore, the industrial union, in many modern industries, is functionally more consistent with the social structure of the industries than is the craft type of union.

Collective action gives the operatives a bargaining power with the supervisory group which they no longer possess as individual workers. The shoe worker's sense of security is enhanced by the knowledge that he and his kind are firmly banded together to defend their common interests against the conflicting interests of the owner-management group.

Membership in a labor union also gives the shoe operative an opportunity to compete for prestige by election or appointment to office in the union. Unions have their own hierarchies of officials chosen from among the union members. Those who rise high in the union hierarchy achieve power and prestige which may rival or surpass those of the top executives in the supervisory hierarchy of the business itself.

Unions foster working solidarity among the operatives and are at the same time derived from such solidarity and the accompanying recognition by workers of their common interests in opposition to the supervisory group. Another effect of social solidarities developed among shoe operatives through their working relations is exhibited in the variation between their participation in the general social life of the community and that of the total adult population of comparable social stratification. Our comparisons showed, for example, that lower-middle shoe operatives tended to a greater extent than total lower-middle adults to associate with upper-lower and lower-lower and tended to a correspondingly smaller extent to associate with individuals who stratified above lower-middle. This indicates that lower-middle shoe operatives were drawn from the "lower fringe" of that social stratum. On the other hand, lower-lower shoe operatives associated with upper-lower and lower-middle individuals to a greater extent than did members of the lower-lower class generally. Thus shoe operatives of the three lower classes seem to form a special segment of those classes, displaying a tendency toward what may be called an upper-lower-class orientation. This specialized shoe-operative behavior in the total social life of the community implies that the working solidarity developed in the factory affects the behavior of the operatives outside of working hours.

There are, on the other hand, numbers of minor conflicts among operatives in the factory which oppose the forces of working solidarity. Our analysis of shoe-operative personnel in terms of sex, age, ethnicity, and social stratification enabled us to interpret many of these minor conflicts and to show that they are often derived from evaluations developed among the operatives in the course of social life outside the factory. Ethnic prejudices, for example, are in many cases powerful forces working against the social solidarity of workers in the factory. (See Appendices 5 and 6.) The ethnic prejudices of the Riverbrookers in the wood-heeling department were so strong that they would not tolerate members of certain ethnic groups in that department. Such attitudes as these, carried over into the factory from the social life outside, tend to disrupt the solidarity of workers.

Our summary has been phrased, thus far, entirely in terms of the community as the "total society." Obviously, Yankee City is in turn but a segment of the larger "total society" that is modern America. In the next few paragraphs we shall point out some of the salient features of the industrial aspect of the integration of Yankee City into the larger society.

In the days of its maritime glory, Yankee City experienced a sense of security and independence that derived from the community's effective control over its own commercial destiny. Yankee City lost its sea trade and had to turn landward for its livelihood. Inland cities had advanced in industrial development while Yankee City was facing seaward, and they had developed natural advantages, such as water power, which Yankee City did not have. The industrial development of Yankee City has been one of trial and frequently of error. The town has been more of a follower than a leader; it has become increasingly dependent upon larger cities; and it has lost much of its independence in spite of its struggles to maintain it. Two developments which have contributed largely to Yankee City's loss of economic independence, especially in recent years, have been what we have characterized as the vertical and the horizontal extensions of industry.

The vertical extension, which results from the growth of a manufacturing enterprise or the merging of several independent factories, creates greater social distance between the higher executives, on the one hand, and the technological workers and factory community generally, on the other. To social distance is often added spatial distance. Both tend to accentuate the split in interests between the superordinate and the subordinate groups in the business enterprise. They also impair the community's control over the local factory and its policies. For when a factory is locally owned the owner and supervisory staff are likely to be influenced in many of their decisions regarding factory policy by the broader community values to which they subscribe and by the fact that many of their employees are old acquaintances and friends. But when a factory is absentee-controlled few or none of the workers are even known to the higher officials and the latter do not feel the pressures that a local owner would feel to conform with the values of the community. Hence the absentee official can set factory policies more nearly in strict accord with the profit-making logic than a local owner, and to that extent the community loses control over the factory and becomes dependent on outside influence.

The horizontal extension of industry has somewhat similar effects on the independence of a community like Yankee City. Trade association findings or agreements which affect factory operations come to Yankee City factories indirectly through the higher officials of the particular enterprises that own the local shoe factories and are indistinguishable, in their effects on Yankee City, from the effects of the vertical extension of industry. But the functioning of a labor union with units in many cities, all directed from somewhere outside the local community, introduces further dependencies of the local community on outside influence. Orders to strike, for example, may come to the local workers from the distant union headquarters, and the local community may be powerless to contravene the order. When there is no inter-community union, social pressure by the community on the workers and/or the factory management may prevent strikes or shorten their duration.

There is an even more basic factor than those already discussed which threatens the security and independence of technological workers and of the communities, like Yankee City, in which they live. This is technological change itself, which is constantly creating new problems for the factory and the community. These problems frequently can be met only by changes in social organization, sometimes minute, sometimes almost cataclysmic. For instance, changes are constantly being made in the techniques of shoemaking, primarily through the invention of new machines which require new operating processes. Such changes always require some readjustment of the social organization of the factories in which the new machines are installed. Frequently such machines make it possible to reduce the number of workers in a given factory and at the same time increase either (or both) the quantity or quality of the product without increasing costs. Under modern competitive conditions the higher executives of the industrial enterprise must frequently act in such situations strictly in accordance with business logic; if they do not, their competitive position will suffer. They cannot solve the problem of providing new jobs for the workers who are discharged as a result of the technological change. The individual community such as Yankee City, particularly in depression years, may likewise be unable to solve the problem of social reorganization that arises in such a situation. This is a problem whose ramifications extend far beyond the confines of any one factory or of any individual community. It affects the economy and social organization of the whole country and should be examined on a nation-wide basis. Individual communities like Yankee City are powerless to solve, by independent action, the problems arising from this source.

X

BLUE PRINT OF TOMORROW—GENERAL CONCLUSIONS

1. *Economic Change and Social Class in America*

THE growing conflict between the opposing forces of managers and workers in the United States is causing intense and widespread anxiety. All that America is, and can be, is irrevocably committed to the outcome of this struggle. Each side, knowingly or not, strives to dominate the basic life and thought of our society. Each seeks to substitute its own beliefs for the general ideology and thereby control the accepted thought of our country.

The opposing forces are the most powerful, intensely energized, and systematically organized movements in contemporary America. Despite the continuing conflict, workers and managers collaborate sufficiently in common productive and other social enterprises to maintain the foundations of America as a wealthy and powerful nation. Furthermore, the most superficial survey of the relations between managers and workers reveals that they are not always hostile; their collaboration ranges from bloody conflict to peace and amity. There is substantial evidence, however, to demonstrate clearly that the changing relations of workers and managers are moving in one direction and are correlated with other changes occurring in the economic and social hierarchies of the United States. Furthermore, it seems possible that we can roughly chart the course of this great movement. If this is true and we can conceptualize and predict the general direction of this movement, it may be possible that we can link action with knowledge and start learning how to control our destinies in order to prevent the destruction of our hopes for a better world.

The Yankee City research was designed to gain insight into the larger problems of American life. In this chapter we shall draw certain inferences from the Yankee City study which we believe illuminate our understanding of the place of industry and industrial strife in the larger context.

The evidence from Yankee City on the decreasing hopes of workers to rise in the skill hierarchy illustrates a phenomenon which is found throughout industrial America. Not every industry has reduced the skilled jobs to the same extent as the shoe industry. In the railroad industry, for example, workers retain their skills; but on the whole there has been a steady decline of skilled jobs and a decrease in the workers' chances to get ahead. Social mobility is no longer present, and the American worker realizes it. The basic precepts of the American creed are thus flouted since Americans believe that everyone has a right to get ahead if he works hard and applies his God-given talents and skills.

It can be assumed that this blocking of mobility in industry should result in strengthening the two great opposing forces, separating them further, and increasing the number of clashes between them. The frustrations of "ambitious workers" trying to rise in the world and take their families with them are the source of common grievance against those above. The decreasing sense of worth and significance on the job felt by all workers adds to this feeling of being stopped by someone or something which is against them. Part of the great strength of labor unions in the United States can be traced to these factors. The unions act as agents to express the common hopes —and bitterness—of the workers.

As the unions have developed and combined into larger units many have become more than weapons for the workers to defend themselves and attack their enemies; they have become social institutions with a great variety of activities which provide for almost every interest of their members. They play an active role in national and local politics. They have developed elaborate social-service programs, health and educational services. In short, they are involved in most of the secular activities of the community. These tendencies are based on the principles and feelings that all workers are alike and are accepting the status of workers who are irrevocably opposed to management, as contrasted with the older principles that differences between crafts are important and social mobility and job advancement certain and uppermost in the minds of the workers.

With such strong evidence from many sources to indicate that the worker group is becoming less mobile and is creating a class of its own with its own social machinery and its own hierarchy in which the ambitious rise to places of leadership and power, it could be supposed that it would be but a matter of time until the groups of managers and owners and those of labor would become completely separated and social revolution would result.

But certain other factors are at work in American society. Mobility in our class system for the worker and his family through advance in the skill hierarchy has never been the only route upward. There have been others, the principal one being education. The American school system has been organized largely to teach people the necessary skills to get ahead. The mature worker's mobility may be blocked at the present time, but his children's chances to rise (he believes) are still good. Once again the American precept of working hard, applying oneself, and learning what one needs to know to get ahead are the guiding codes for the worker and his child. The child can climb upward on the several rungs of the primary, secondary, and college levels and reach positions of power and prestige in the ranks of industry, business, and other social hierarchies. The grammar and high schools, the state universities, and land-grant colleges have provided an inexpensive and available ladder for the ambitious. As long as this route remains open to the children of workers and other members of the lower class, their frustrations will not be sufficient to be explosive.

The chart called "Social Mobility in the School; Blocked Mobility in the Factory" (Chart X) graphically represents some of the more significant aspects of social mobility in contemporary America as it affects the worker. The longer rectangles at the left and right represent machine and commodity factories, and the central rectangle, the public school system. Management is placed at the top of the industrial hierarchy and the workers at the bottom. The small rectangles labeled "machines," suspended from "management," with their arrows pointing downward to the double line which separates the worker level from the managerial, represent the blocked mobility of the worker. The downward-pointing arrows tell the story of thwarted worker-mobility in the factory. The arrows running from each of the outer rectangles into the rectangle labeled school, which point upward and connect with the other arrows within that rectangle and ultimately return again into the factory rectangles at the top levels, tell the story of upward mobility through the school hierarchy into the top levels of industry.

The school is taking the place of the factory for the mobile and ambitious children of workers. The pressure on public and private schools to provide courses which will train men and

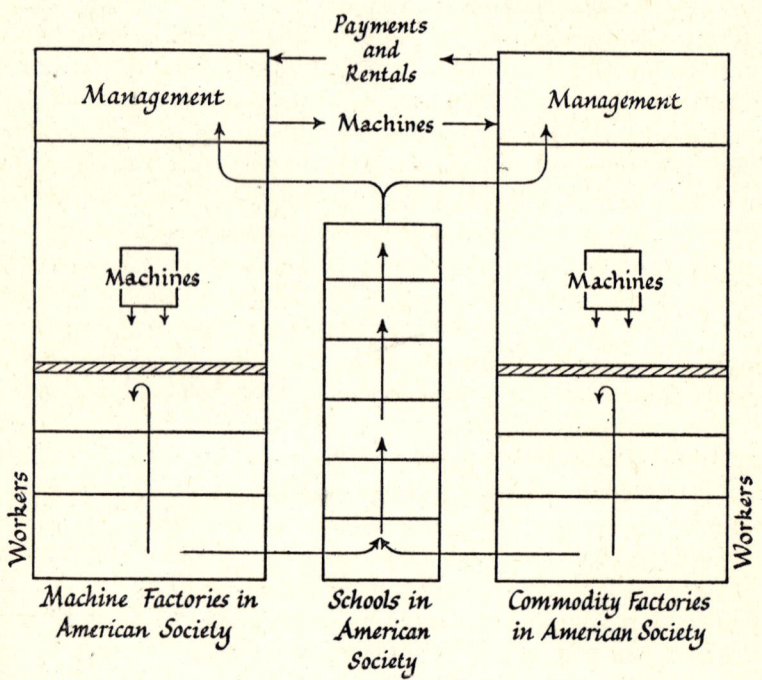

Chart X. Social Mobility in the School; Blocked Mobility in the Factory

women for the new professions, for other less highly rated occupations, and for the older and more highly rated professions, is another indication of how the people, particularly the workers, are using our schools. Many large industries have half-consciously recognized the change and are establishing their own schools where basic training as well as special training is provided for the worker. The unions, too, partly to maintain control over the sentiments of the workers, are recognizing the powerful appeal the school has for the young ambitious worker by establishing their own schools to educate the men in general subjects and train them for leadership under the auspices of the union. The rise of the great correspondence schools is in part attributable to this change in our mobility process. Their appeal is to offer a royal road of education as the easy route to success. From a careful scrutiny of a number of communities, life careers, and occupations, the authors believe that education has become the preëminent and most-used route to power and prestige for all classes of men in American society.

At first sight one would suppose that this tremendous development of educational facilities would increase the opportunities for mobility and decrease the chance for revolutionary outbreaks expressing frustrated aspirations. Certainly this assumption is partly true and a significant chapter in the story of the changed mobility patterns in our class order, but another vital factor must be considered. The evidence from Yankee City and other places in the United States strongly indicates that mobility through the schools is also slowing up and that the higher positions tend to be filled in each succeeding generation by the sons and daughters of families who already enjoy high positions. The evidence from a great variety of studies clearly demonstrates the truth of this last statement.[1] While newer educational routes are being formed for the ambitious, the older ones are becoming increasingly tight, and it seems predictable that in time education may not be a certain route for those who seek success. It seems probable that our class system is becoming less open and mobility increasingly difficult for those at the bottom of the social heap.

Until now we have considered the predicament of the American working man in the lower classes. It might be supposed that with a tightening mobility system the positions of more advantaged people would be increasingly secure. This assumption is certainly not true. The upper-middle class is an unhappy group, beset economically by increasing taxes which have low-

---

1. See W. Lloyd Warner and Paul S. Lunt, *The Social Life of A Modern Community*, "Yankee City Series," Volume I; and Robert J. Havighurst, W. Lloyd Warner, and Martin B. Loeb, *Who Shall Be Educated?* (New York and London, Harper & Bros., 1944).

ered the incomes of the business man and decreased the real income of those who live on fixed salaries and professional fees. Subsequent to World War II, many of these people have lost confidence in their own ideology of success and hard work and have become more and more fearful of the increasing strength of the organized masses. Being a small group, most of them feel that our democratic political structure does not represent their interests. Those in the small cities and towns know that their control over their own economic and social destinies has been lost to the men of the big cities or to growing controls from Washington.

The upper class of America is a very divided group. Many of them depend on stocks and bonds for their entire income; the income from these investments is dependent upon the financiers of New York, London, and other international financial centers. The economic interests of the upper-class families in their local communities are disappearing. Some of them, in response to the shift of economic interest, have left the smaller cities and gone to the great metropolises, leaving many of the smaller cities without leadership from the people who have traditionally supplied it. Others have stayed in their local communities to retain their social positions as members of "old families" in the local lineages—positions which would be lost in a large city or new community.

The upper-class families who stay at home are more often from the older and more settled regions of the country. Those who quit the local communities are frequently from the Middlewest and newer regions. The New Englanders and southern old families possess bigger and more ancient social "investments" in their family traditions. Being more highly evaluated by their local townsmen, and for that matter by themselves, such people feel more highly rewarded in their local communities than in those with only the very recent background of the Middlewest. However, many of the sons of upper-class families in the South and New England do not return to their own communities after going through college but often establish themselves as officials in industries and banks in the big cities. Possibly it is not pure chance that many of them enter business organizations that once had their headquarters in the smaller cities. It is significant that few of the Yankee City shoe owners were higher than upper-middle class. A generation ago the owners of these factories would have been either in the upper class or securely on their way up to it. Power, money, and prestige have flowed up and out from the small city to the great metropolises.

The economic relations among members of our middle and upper classes continue shifting. Their acquisitive and competitive motivations and the means of expressing them are increasingly curbed. The members of the upper and middle classes now express feelings of deprivation that were once characteristic only of the workers. The federal government as well as many of the state governments has invaded what was once the domain of private business and, by regulation and the force of public opinion, greatly limited the area of free choice for the business agent. The men of top management now feel the loss of power and freedom of action. The great unions have further limited their freedom.

Meanwhile, their own trade associations, by developing agreements among business competitors which increasingly narrow the area of competition and regulate and restrain the freedom of each of the members, further constrict the managers' freedom, indicating that the internal organization of business is undergoing changes which are decreasing free enterprise. The so-called free-contract characteristic of the relations of labor and manager, in which the individual worker agreed to sell his services for a given wage, has largely vanished. The concept of the "free contract," when exercised by management, was a powerful protector of the economic position of the upper-middle and upper classes and helped control the actions and demands of the lower-class workers. The government now defines the minimum wage; social security as well as other legislative stipulations makes the "free contract" less and less important in the field of labor-management relations.

The structure of capitalism is no longer free; it is a regulated capitalism. The big-city capitalist, admired and envied by those of the small town, is not a free man, and he is not the free capitalist that his small-city predecessor once was. It is abundantly clear that the squeeze felt by all those in the higher income brackets and social classes is not only economic; in addition their feeling of worth, prestige, and significance has been severely reduced. The emotional tone found in the editorials of their newspapers, magazines, and the speeches of their politicians are symbolic symptoms of their gathering sense of lost rights and privileges.

Despite the great changes in our economic and social order, the American class system continues. When the economic relations shift we expect changes in the status system. There are many who believe the increasing power exercised by the political, economic, and associational organizations of the workers will result in an equalitarian society. The present writers do not share this opinion. We believe that, with the increasing complexity of our society and the greater development of our relations with other great world societies, the hierarchical forces operating in our society will increase their power. Our class system is changing and will continue to change, as will other systems of our society, but such changes are likely to strengthen rather than weaken our rank orders. Obviously an increase in the distribution of goods throughout the society, resulting in more consumption for the lower classes and general betterment of their condition, will modify the relations of the lower, middle, and upper classes. The development of new occupations and positions of power in labor and governmental hierarchies will change the relations of economic status and social rank. The rise of new industries and the decline of others will contribute their share to the shifting of social and economic rank relations. The diminution of prestige and economic power suffered by the western powers in Asia will contribute indirectly to changes in the status relations of our own society.

Despite all of these and other significant modifications of our economic power, our basic hierarchical social order will remain. Men and women will continue to strive for the higher rungs in the ladders of status because prestige, power, and greater rewards are always at the top. This statement is true of the church, schools, and other systems. A cardinal will out-rank a bishop and both of them will receive far greater rewards than the parish priest or local minister. A superintendent will outshine a grade-school teacher and a Supreme Court judge will stay on a distant pinnacle above the lowly justice of the peace.

No complex society can escape such ranking and get its work done. Collaboration among men is impossible without group leaders. Collaboration between groups is impossible without the aid of higher leaders to coördinate their efforts. The diverse groups of each level must depend upon the level above to help them coördinate their efforts. The more complex the society the more numerous the strata and the greater the social distance from the top to the bottom. The problem of coördination at each higher level increases enormously since the diversity of interest, social characteristic, and the form of opposition greatly increase among the groups whose activities below must be organized, integrated, and directed.

Those who occupy relatively similar points in the several hierarchies of a society tend to participate at the same social levels. Their children tend to marry at their own level and to be trained in the social values and customs developed by their own sub-culture. This principle holds as much for Soviet Russia and its equalitarian dogmas as it does for the United States with its democratic theories of human equality. One of the principal differences between the two is that the passage of time has given the people at the top in our society opportunity to solidify their position. It has also given such members of the society more time to define what the top position is.

Economic and social change may greatly modify our status order, new names may be given to the top and bottom, and new kinds of people may occupy these positions, but essentially the same basic relations of power and prestige will continue and, from all present indications, increase in strength; and the social distance between the top and bottom will be extended.

2. Species Behavior and a Planned Society

THE destruction of the skill hierarchy is not peculiar to Yankee City; it is a characteristic of changing conditions in American industry generally. Paradoxically enough, as jobs become increasingly specialized and divided into smaller, composite units, they become more and more alike because the simplified operations and knowledge required in them are reduced to a series of repetitive actions which have a larger number of common elements. Several important consequences derive from this process of specialization: the learning task in acquiring the "skill" is easy, the period necessary to acquire proficiency short, and, once a job has been simplified and routinized, its processes can be incorporated into the actions of a machine and the machine substituted for workers. The activities of the workers who run the machines are then further simplified and routinized. The skill once "owned" by the worker and sold as a service is now possessed by the manager in the form of the machine.

As an industrial hierarchy increases the distance between the workers and management, the human being who is a worker becomes a digit in a vast series to the man at the top. Likewise, the administrator ceases to be a man and becomes a symbol of economic authority to the worker. Each tends to treat the other in terms of this over-simplified definition. Mr. Cohen from New York saw Sam Green and the other hundreds of human beings in his Yankee City factory as work units and as digits in an employee total. Sam saw Mr. Cohen as the boss for whom he worked and who gave him his wages. Each saw the other also as a symbol of a region and a culture but, above all, as a symbol of a status. Each had reduced the other to a non-human level. The chain of relations which connected them had been reduced to the cold exchange of wages and services in the precise relations of wage earners and managers.

Technological advances in older industries often appear as part of a mopping-up movement which follows rather than precedes the specialization of jobs. "New" industries, using the experience of the older ones, break their operations into simple units and later mechanize them; they do not go through the evolutionary processes of the older industries. The experiences of the new radio and airplane industries give clear evidence in proof of this point. The theory that technological advance always precedes change in the social organization and that there is a "lag" between the forward steps of a technology and the laggard ones of a social organization seems open to serious question. Some writers believe that the course of progress of western civilization may be charted by change in the social structure following technological advance. Our evidence indicates that this is only a partial truth. It is equally tenable that the increasing division of economic labor, with the division of jobs into smaller and smaller units and their combination into larger and larger economic structures, is more often than not the forerunner of mechanization and technological advance.

The segmentation of the daily life of the American community and its increasing complexity (accompanied by a corresponding movement to reclassify the social diversities into larger and larger comprehensive social units) have resulted in the field of economic behavior in great holding companies and the gigantic cartels which relate and integrate myriads of diverse smaller units into common corporate enterprises. Meanwhile, the control over these great structures moves steadily up an ever-extended hierarchy into the hands of a smaller and smaller number of powerful men. In the area of government, local grass-root activities become more diverse and more specialized while the larger political units at higher levels continue to absorb power and control. But power inevitably moves into the upper reaches of the governmental ladder. Associations also are becoming increasingly diverse in interest and purpose while their memberships develop larger "over-all" organizations which combine the diverse local units into great community and national enterprises. The philanthropic associations, for example, coördinate their activities into fund-raising and "educational" campaigns and set up permanent machinery to maintain their common efforts; special-interest and propaganda groups follow similar tactics and develop social structures of vast design and unmeasured power.

A greater and greater simplification of activities accompanies the division of the social body into an increasing number of "organs," and the latter into more numerous cells, which are subsequently related to the larger national society. At the level of the job this process results in specialization. At higher levels it is expressed in increasingly explicit, logical thinking about what is being done and the relationships between jobs. At the very top levels there is the effort not only to analyze logically and scientifically what is being, and should be, done but also to work out predictable behavior for the whole organization in terms of its known needs. Meanwhile the content of behavior from the lowest levels of the organization to the executive office is increasingly subject to scientific control.

Planning for an industry, a government, an association, or a whole society is but the ultimate expression of rationalization of behavior at all levels of our social hierarchies. Planning, of course, is an attempt to force human behavior into a more logical and explicitly recognized and desired mode of action while trying to project such behavior beyond the present into a secure future. Such things as accident, bad luck, and unexplainable failures are thereby guarded against; and that highest of all virtues recognized by a capitalistic or any other human society—the reduction of risk-taking—is accordingly emphasized. It should at least be remembered that when risk-taking is reduced to a minimum in the economic and other affairs of life, and daily existence becomes more secure, other forms of risk-taking which are now highly approved by almost everyone tend to disappear. That which we call "adventure," the thrill of "the hundred-to-one shot," or "hitting the jackpot" are correspondingly reduced and may become socially disapproved. But it is doubtful if any society, no matter how logically and systematically it is fashioned, can escape maintaining risk situations which will be defined as socially desirable by the group.

This brings us to what is called the "human factor" in industry and other highly rationalized institutional systems. A human social maze may be so constituted that the planned society can train its citizenry from infancy to learn to act as rational men in a rational social universe. But these same men and women are composed of muscle and bone, viscera and nerves, and animal impulses. As members of their species they possess a sub-cultural interactive system of behavior which is species behavior, which existed before culture and still lives within it. This basic species behavior must forever be expressed through cultural forms and seep into the rational and logical world of men; thus the world of human feelings and of deep sentiment must forever establish its place in men's actions and their social systems.

Risk-taking has been decreased enormously in man's relations to nature. The margin of safety has steadily increased, and some small advance has been made in understanding and controlling the human factors to reduce risk in human interaction; but no people can impose logical and rational action on its social order too far without experiencing serious and possibly disastrous results. All too often in the minds of the planners, planning means the reduction of the impulse and non-rational life of the people and the increase of logical behavior. Successful planning must necessarily be fashioned around the basic needs of our human kind as members of an animal species.

The family in large part provides the institutional control which men have fashioned for this purpose. Behind its walls impulse life is permitted. It can be argued that our present and past efforts to reduce economic risks and to rationalize social relations have reduced the role of the family and enormously decreased its effectiveness as a place where the two sexes and the two generations can satisfy the deep basic needs of their species and discharge their animal energy in a form suitable to them and their society. When built on the fragile and explosive foundations of family and community systems in which the basic human needs of the people are not satisfied and often frustrated, such well meant schemes will always be failures unless those who fashion these logically planned orders take account of the species whose culture needs must be cared for. No social order extending itself vertically and horizontally, while systematizing and rationalizing the actions and relations of its members, can ever achieve any kind of permanent equilibrium.

3. World Implications of the Extension of the Economic Hierarchy

WHILE our corporate hierarchies have heightened and expanded into more powerful enterprises and spread throughout the nation and the world, the unions have developed from weak, ephemeral, and loosely organized local units into vast systems with their own powerful hierarchies whose higher officials deal at a coördinate level with top management. The effects of these developments cannot be overemphasized. Not only the executives of local factories and unions lose the power to deal with each other as they did in Yankee City, but the executive heads of great companies are now relinquishing their right to make decisions in favor of the representatives of the whole industry who meet and deal with the union representatives of workers from that same industry. Although their common industrial interests are basic, the status interests of the top levels of management and the workers are in opposition. This conflict necessarily results in the employment of a third party to act as a referee. Since the church has lost its great power as a regulator, the only referee now available to the high representatives of these great opposing forces is the government. As the other hierarchies have elaborated larger and more diverse units into new structures, the government has reduced the power of its local and secondary offices and placed power in the upper brackets of the federal state. The referee between the two great conflicting forces of workers and management must necessarily be endowed with great social power and be able to apply sanctions with sufficient force to maintain moderate collaboration between the two for the good of all the people.

As long as the governmental, labor, and management systems remain separate, the partial or full integration of American society can only be maintained by the subjugation of one of the contestants by the other, by the destruction of the power of one of them by the government, or by the government's treating the whole matter as a combat between two equal forces where each side can score a little but not too much. This latter condition will probably remain with us in America until a time when our country's social system re-integrates itself and the opposing forces now focused in the conflict of capital and labor are controlled or express themselves in some other form.

There are increasing evidences that, if catastrophe does not overtake us, the power to compose labor difficulties will move beyond national governments to an international governmental locus. The present chaotic condition of world affairs is not so much due to ideological differences, great as they may be, as to the failure of human beings to evolve a substantial social organization which will encourage and reward coöperation and collaboration around the common enterprise of making the world socially inhabitable. The failure to fashion such a system should not necessarily be made a moral score against us, for it is possible that the processes to which we have referred earlier in this chapter may be operative but not apparent to us because we are too close to what is happening. It may be that the dual and interdependent processes of specialization and greater division of labor, on the one hand, and increasing social complexity, development of larger groupings, involvement of a greater number of people, and lengthening of the social hierarchy, on the other, are occurring with almost revolutionary speed. Until now these processes and the social movements which ideologically express them have been centered in the great world powers. The British Empire has built one system, Soviet Russia another, and the United States a third, while China is desperately attempting to construct another. Japan and Germany—temporarily, at least—have failed in their attempts. Elaborate economic, political, associational, and ecclesiastical hierarchies have been established in each of these areas to form the scaffolding of cultural activities. Meanwhile the differentiating processes taking place in each of these countries have made the citizens more alike, while forcing the citizens of each to be aware of the fact that they are members of different systems.

The structural hierarchies of the separate world areas interact outside their own orbits with those of other regions, but each areal system possesses strong autonomy, and the locus of power is centered within it. Great international capitalistic enterprises, often monopolistic in character, have succeeded in effectively crossing national boundaries and have developed methods of organizing some of the diverse economic units of the world. Opponents of such systems continue to fight what the authors believe to be no more than rearguard battles which only delay the advance. Such capitalistic enterprises are the enemies of nationalism, and, as citadels of capitalistic power, they are the foes of labor and of the remnant forces of nineteenth-century liberalism. The cartel, one of the most powerful forms of international capitalism, must be recognized as a new social structure, developed by us in our desperate efforts to reorganize human behavior to function on an international basis. At the present time such economic institutions may or may not be evil in their effect, but international economic institutions of some kind are absolutely necessary if the world is to evolve a reliable international order.

Technological processes which are already in existence—such as instant world-wide communication, rapid transportation, and international exchange of goods and services—and the social process inherent in the increasing divisions of labor in each of the larger social areas make people everywhere increasingly alike and interdependent. Therefore, it seems probable that, unless disaster forces us to retreat to barbarism, the present international economic systems must, after adjusting to new social controls, ultimately triumph.

As we said earlier in this volume, the extension of institutions over more diverse peoples and activities (horizontally) must be accompanied by vertical extensions which increase the social distance from the bottom to the top of institutional hierarchies. The extension both horizontally and vertically of the international economic institutions over more men and a greater variety of their activities will only be secure when the economic institutions are complemented by church, associational, and political hierarchies. The development of these other institutions will permit them to act internationally as counterforces to achieve a balance of power between them and the economic order, much as our society once enjoyed on a national level when there was equilibrium between the social and economic institutions in Western Europe and in the United States. It is impossible to predict what the new social order which is now evolving will be. Once the society is formed it is doubtful if such labels as "fascist," "communist," or "capitalist" could be realistically applied to it. We can say with certainty, however, if such a system is formed, that the social principles characteristic of hierarchies will be stressed more than at present, since the peoples who compose it will be more diverse and more difficult to organize, and the need for lines of authority and responsibility will be greater than in any other time in the history of man.

APPENDIX 1

THE SHOE FACTORY AS A SOCIAL STRUCTURE

THE social organizations of the seven shoe factories in which the strike took place were varieties of the same kind of socio-economic institutions found in Yankee City and in America generally. They were all hierarchical systems of rank in which the basic relation was that of the employer-employee. Functionally, these systems made shoes to sell to the consumer, to make profits for the owner and wages for the worker.

The internal and external organization of the American factory is depicted in Chart XI, a schematic representation of the four major segments of a large shoe-manufacturing enterprise. It indicates the relations of these segments to one another and to the public. The heavily outlined triangle at the base of the figure represents the shoe factory in its relations to the rest of the enterprise and to the local factory community.

In our diagram each of the four main substructures of the enterprise is related to specific portions of society. At point A, where capital flows into the organization, that portion of society which has surplus wealth or credit presumably engages in the enterprise because of the opportunity afforded for safeguarding or increasing its wealth.

At the left-hand side of our diagram is the purchasing division, related directly to that portion of society which engages in the production of materials used in shoe manufacturing. The contact may be with the producer of raw materials directly, or indirectly through the medium of the manufacturer of partly processed or finished goods used in the varied processes of the enterprise.

The lower part of the diagram represents the shoe factory and its relations with the community in which it is located. The factory affects the community by giving employment to local people, by the payment of wages and of taxes. The local community does not necessarily profit directly by the total enterprise, produce any of the goods or raw materials used in the factory operations, or consume the factory's products.

The right-hand side of the diagram represents the distribut-

CHART XI
The Formal Structure of a Large Shoe Manufacturing Enterprise

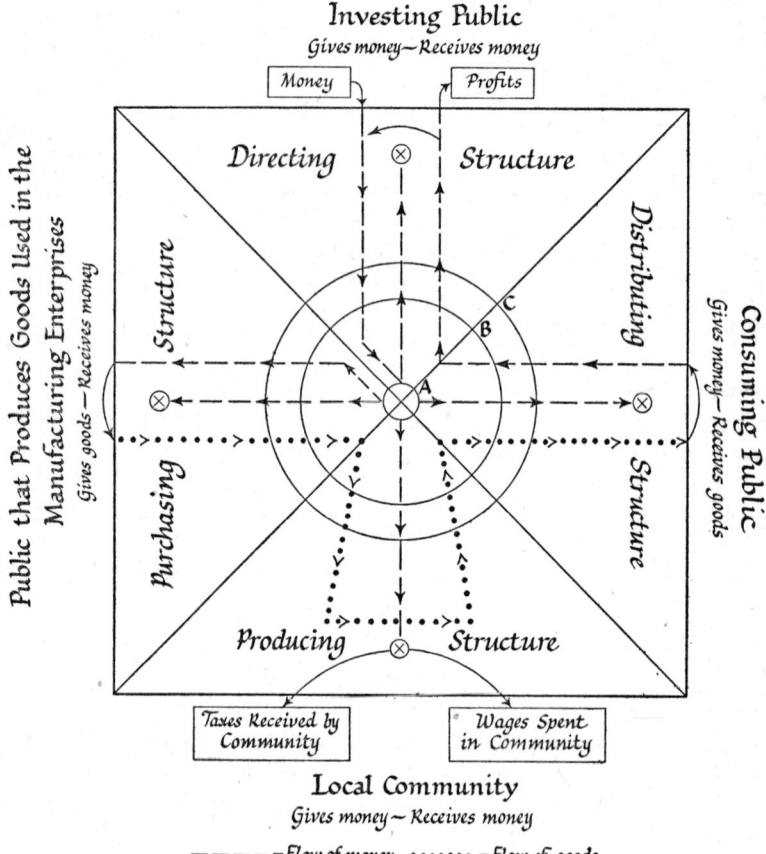

ing substructure through which the shoes made in the factory are sold to the consumers in exchange for money.

The whole business organization, if it is small and if it makes from local materials a product which is specifically adapted to the needs of the community, may affect the local community from each of the four points of our diagram. As organizations increase in size and complexity, however, this condition very seldom occurs. The factory, by reason of the fixity of its plant, is the portion of the general manufacturing organization that most directly affects the community. No matter how complex a business organization may be, a particular factory functions as a concrete point in the life of the community in which it is located. This is true even in the extreme case where all the materials are brought in from the outside; the products are all sold elsewhere; and all the money to operate the enterprise comes from distant sources.

In the case of a large organization operating many manufacturing units (such as that which operated the absentee-owned shoe factory in Yankee City), the local factory is spatially separated from the general purchasing, distributing, and directing structures. Local managers requisition their major supplies from a central purchasing department and ship the finished products to a central distributing organization. They are accountable to a major executive or an administrative body.

This latter executive or controlling group relates the individual shoe factories to the larger society and handles their various distributing, purchasing, legal, and financial problems. The executive has another function—that of coördinating the activities of the various production units through their individual managers.

A smaller organization with only one plant, owned and controlled locally, has a somewhat different structure. Here the function of the major executives merges with that of the factory manager. This executive handles the problems of external relationships as well as those of internal supervisory control. In the earlier days of the one-man shoemaking establishment and the "ten-foot shop," the shoemaker did all his own work and made all outside contacts as well.

APPENDIX 2

THE WORKERS AND THEIR DEPARTMENTS

THE major jobs in the various production departments of a turn-shoe factory and the working relations among individuals in each department are described here.[1] Before doing this, however, we shall discuss the sequence of technological processes and the relations of the various departments to one another in the production of finished shoes.

The three principal parts of a shoe are the sole, the upper, and the heel. Each of these was manufactured separately; the upper was then fastened to the sole, and finally the heel was attached to complete the shoe. Leather, of course, was the most important basic material, for nearly all soles and most uppers were made of it. Calf and kid were most frequently used for uppers, but kangaroo, lizard, alligator, snake, and other skins were also used. Cloth for the lining of uppers was also an important material. Some heels were made of alternate layers of leather and composition, others of wood.

Sole leather was received at the factory already tanned, treated, partly graded, and cut roughly to shape. Only one thickness of sole leather was used in the turn process and it was prepared for the attachment of the uppers in the sole-leather or stock-fitting department.

The leather for uppers also came to the factory tanned, treated, and roughly graded. It was cut into the desired shapes in the cutting department. The cuttings were fitted together and stitched in the stitching department where the linings were also sewn in. From here the uppers went to the assembling department where workers added soles of the proper sizes (received from the sole-leather department), counters, shanks, toe pieces (all made of composition and supplied ready-shaped to the factory), and lasts (wooden molds on which the shoes were built). These separate parts were assembled in case lots

1. There was one small factory using the turn process in 1945.

(usually thirty-six pairs per case) and sent to the making room. Here the uppers were attached to the soles.

Leather heels were purchased by the factory in crudely shaped blocks made of alternate layers of leather and composition glued together. Wooden heels were usually received already finished, covered with leather, cloth, or celluloid. Heels were sent from the stock room to the leather-heeling or wood-heeling department; crude shoes from the making room were sent here also. The heels were attached in the heeling departments; in the leather-heeling department the crude heel blocks were also shaped as desired.

From the heeling departments the shoes were sent to the finishing or bottoming department where the sole received its final finish. From here the finished shoes moved into the packing department for minor cobbling and final inspection. Then they went to the shipping room.

The various departments were so located in the factory that a minimum of time and effort was required in routing the parts through the processes. Although some small factories occupied but a single floor, some of the larger ones occupied buildings four or five stories high. In the latter, the manufacturing processes generally began on the top floor, the shoes progressing downward in the course of manufacture until they emerged from the packing department near the ground floor shipping exit. Special considerations caused some changes in this general routing scheme. The cutting and stitching of uppers, for instance, required the best possible light; therefore, these departments were ordinarily on the top floor. Sole leather must be dampened in order to be worked; on the theory that the air is damper there, this department was usually located on the ground floor.

Some of the buildings which housed Yankee City shoe factories were adapted from textile mills, abandoned when the textile industry moved away. They were not fireproof, but were ordinarily equipped with sprinkler systems and had reasonably adequate facilities for exit in case of fire. Modern artificial lighting made it possible to arrange work benches and machines without reference to windows. The larger factories maintained a first-aid room in charge of a registered nurse.

We shall now proceed to the examination of the jobs[2] and social structure of each major production department.

The Sole-Leather Department

In this department, usually the smallest in a turn-shoe factory, the soles were prepared for attachment to the uppers. In the department we have chosen as typical for concrete comparisons, there were fourteen sole-leather workers, all men. They usually worked in their undershirts and pants. Some type of apron was worn by all shoe operatives.

The sole leather came to the department in burlap bags. It was already cut roughly in the shape of soles, but varied in size and thickness. The first job in the sole-leather department, or stock-fitting department as it was sometimes called, was the sorting and grading of these roughly shaped soles. The sorter was merely a handler of objects, but his work required some knowledge of leather. After the soles were sorted and certain sizes and shapes were prepared for processing, the next operation consisted of "wetting up" the stock. Here the leather was put in a water bath in order to soften it sufficiently for later operations. The "wetting up" process was performed entirely by hand.

The next process was a machine operation in which the fleshy part of the sole was sliced off, the worker merely feeding the material to the machine. In the rounding process the rough sole was placed in the sole form of a machine and cut around the form to the desired size and shape by a mechanical knife. These forms varied according to the specifications for each particular run of shoes.

The sole was passed to a stamper who, by a machine operation, stamped it with a number or symbol representing the size and style. The sole then went to the skiver or splitter, who operated a machine which sliced a thin layer of the outer part of the sole from the heel to the shank. This sliced piece was left attached to the shank of the sole and was later used in the wood-heel department to cover the surface of the front part of the heel. Here again the operative merely fed the sole to the machine.

---

2. The description of jobs, although necessarily superficial, will suffice to indicate the general types of technology used.

The most difficult machine job in the sole-leather department was that of the channeler. His task was to feed the sole into a cutter machine in such a way that a channel was cut around the edge of the sole on its inner side. It was through this channel that the Goodyear stitcher in the making department later sewed the sole to the upper. Channeling required a considerable amount of training and precision, as the channel had to be cut in a definitely prescribed section of the sole. The technique of the operative lay primarily in the proper feeding of the material into the machine.

The next process, called "marking for shanking out," consisted of drawing a line on the sole to aid the shanker-out in trimming the shank of the sole. This shaped it more accurately for the later operations. Machines were used in both of these processes. The last operation in the department was the application by hand of a stain and cement to harden the outer edges of the sole. Then the sole, ready for attachment to the upper, was sent to the assembling department.

The Cutting Department

The eighty male workers in the cutting room performed the first operations on the materials for uppers. Hides came to them already sorted for texture and color. The number, shapes, and sizes of the cuttings each operative had to produce from a skin of a given size were determined in the business office. Leather for uppers was light enough in weight so that it could be handled and cut without being moistened; therefore, there was no need for this department to be located on the ground floor. The cutters had to exercise care in order to avoid injuring the leather, causing blemishes or other defects in the finished shoes. Consequently, cutting operations required the best light obtainable. For this reason the cutting department was usually situated on the top floor of the factory.

Three types of operations, all involving the use of knives, were performed in the cutting department. They were called machine cutting, hand cutting, and hand dinking. Each will be described in turn.

The techniques of the machine cutter of outside leather were those of both object-handler and machine operator. The machine used was known as a clicking machine. It was very heavy and rested directly on the floor. The operative stood before a hard rubber composition cutting board attached to the machine, and placed on it the leather to be cut. In the clicking machine, the knife was a heavy metal die with cutting edges on top and bottom. The operative was furnished several of these dies. He selected the one required for a particular operation and placed it lightly on the leather. This necessitated great care, as the sharp cutting edge of the die might easily mark the leather. The die had to be placed in such a way that, in the completed cutting, the grain of leather would run in the proper direction considering the strain eventually to be put upon the shoe. The cutter also had to use care in matching the color and shade of the leather for similar cuttings for each shoe of a pair.

A heavy metal arm with a flat under-surface swung horizontally over the cutting board. The operative guided this arm over the die and, when it was in position, released it by means of a lever. The arm dropped vertically upon the die and forced it through the leather. It then automatically returned to its former position. This arm had to be extremely heavy in order to force a die through a piece of leather; it was because of this arm that the clicking machine was itself heavy and bulky.

If the operative wanted to make the same cutting for the other shoe of a pair, he removed the die and reversed it or else selected a different die to make another of the prescribed cuttings. He tied together the cuttings for each pair of shoes. When the outside cuttings for a case were completed, they were placed on a rack, with the lining and trimming cuttings, to be sent to the stitching department.

Since styles in women's shoes change so frequently and rush orders are so common, it was usually profitable for a factory to maintain a staff of hand cutters of outside leather. The hand cutters stood at wooden benches, each of which was supplied with a cutting board to hold the material. The die used in this work was of cardboard with metal edges. The operative placed the die on the material and, with a sharp hand knife, cut the leather around the outer edge of the die. He was as careful as the machine cutter in properly placing the die on the leather, although the sharp edge of the heavy die used in the clicking machine required greater care in adjustment. Hand cutting, like machine cutting, necessitated a knowledge of leather. However, more technical experience was required in the hand operation, for it contained a greater chance of error.

In the cutting of linings both hand and machine methods were used. It was customary to cut many pieces at the same time. Hand cutting often required greater physical strength than cutting by machine.

Some of the smaller and less important cuttings of leather, particularly the smaller cuttings of linings, were made by what is called the hand-dinking process. Here the cutter, standing at a bench, adjusted his material on the cutting board and placed upon it a heavy metal die with a cutting edge on the bottom. The die was perhaps three inches thick with a heavy metal post in the center. The operative held this die in position with his left hand, holding a hammer in his right. By hitting the post with the hammer, he forced the die through the material. The operative cut many pieces before he had to remove them from the die, thus saving considerable time. Very little practice was required for proficiency in this operation.

These three methods of cutting were used concurrently in most Yankee City shoe factories. It has been estimated that unless a factory had an order for at least 250 cases of a certain type of shoe, to operate with hand cutters was cheaper than with high-priced machine dies. In large factories, however, there was less need for hand cutters, for the normal order was of sufficient size to warrant the purchasing of clicking-machine dies.

Cutters are traditionally the aristocrats of the shoe trade. In the old days it is said "they came to work in high silk hats." One reason for such high status was that their work largely determined the quality of the finished shoes. Furthermore, years of training were necessary before a man could know leather as did the old-time cutters. Now, however, leather is much better sorted before it reaches the cutters and they no longer need extensive knowledge of leather. The cutter's job has always been one of the cleanest in the factory. Cutters even in recent years have worn shirts and neckties under their work aprons; almost everywhere else in the factory the men remove their shirts before going to work and wear the apron over their undershirts.

There appears to be little reason other than tradition—unless it might be cleanliness of the work—for cutters to feel superior to other workers. As a man of long experience in the shoe trade said: "Although cutters still seem to feel superior in a shoe factory, this superiority does not evidence itself much outside of work. There is certainly a definite leveling process at work in the industry."

The Stitching Department

In any shoe factory using the turn process, at least 30 per cent of the operatives would probably be stitchers, especially if the factory made novelty shoes. This was true of our typical factory where nearly a third of the operatives—313 out of 985—worked in the stitching department. Three-hundred and six of the stitchers were women; only seven were men.

This department was primarily concerned with fitting and sewing together the uppers and linings which had been received from the cutting room. In addition, various operations required by different styles and designs were performed on the cut pieces of outside leather. Many of the fancier novelty shoes required decorative perforations in the uppers and a great deal of fancy stitching. One pair of novelty shoes might necessitate as many as sixty or more operations in the stitching department. The accessories used in this department were commonly known as "findings."

Jobs in the stitching department, like those of the cutters, were among the cleanest in a turn-shoe factory. The materials were dry and had had little previous handling. The physical labor was lighter than in most other work in the factory, but great concentration was required. The emphasis on speed, everywhere apparent in a shoe factory, was especially noticeable in the stitching department. Although the workers sat close to one another, each had to be so intent on his or her work that conversation was precluded. The silence of the stitchers at work contrasted decidedly with the hum of conversation characteristic of some other departments. The only noise—surprisingly little—came from the machinery in the room. The various electric machines of a modern stitching department were supported on tables which usually accommodated ten or a dozen workers, five or six on each side.

We cannot describe within reasonable space more than the five most important [3] jobs in the stitching department. One of these was skiving, a machine process by which the edges of the vamps and quarters were beveled or skived. The edges were applied by hand to a revolving knife and thinned so that the pieces could be sewed together in uniform thickness.

A second group, the fancy stitchers, did the decorative stitching on the outside of the uppers. Sometimes this stitching also had a useful purpose in the construction of the upper. The operatives used a machine with a flat sewing surface on which the cuttings were laid and guided to the needle. The stitching followed lines marked on the upper by a previous machine operation. Different styles of fancy stitching demanded different types of machines, but in general the technique required was the same for all. Some of the more complicated machines had two, sometimes four, needles which operated at the same time to make double or quadruple lines of stitching.

The vampers sewed the vamps to the quarters after fitting and adjusting the skived edges of each. The sewing was done on a machine resembling the fancy stitcher. The linings which had also been skived were stitched to the uppers at this time.

The top stitchers, using a machine in which the needle operated on a post rather than a flat sewing surface, sewed around the top of the shoe neatly binding the lining to the outer leather. The French-cord operatives cemented one edge of the narrow strip of French cord on the inner side of the top of the shoe then turned the loose edge over the top and carefully sewed it, making the final finish of the top part.

Although these five processes were considered the most important in the stitching room, they are merely selected examples of the work performed there. The number of operations carried on in this room in making one pair of shoes varied greatly according to the type and style of shoe. The uppers, after this department had completed its work, were sent to the assembling department where all the essential parts of the shoes were gathered.

3. Composite judgment of a stitching-room foreman, an executive of the union, and a woman stitcher who had had many years' experience.

The Assembling Department

To the workers in the assembling room (fifteen men) came the finished uppers from the stitching department, the processed soles from the sole-leather department, and the lasts, counters, shanks, and toe pieces from the various stock rooms. The assemblers collected, sorted, and arranged these items in sections of rolling crates, each section containing sufficient articles to permit the making department to turn out one case of shoes. The articles were assembled according to specification tags which listed the assortment of sizes for a particular case of shoes. The assembling department usually delivered the crate with its assembled materials to the making room.

The workers in this department were simply object-handlers; their work was analogous to the sorting and assembling of materials performed in any factory. They were not engaged in processing materials and were not, in the strictest sense, shoe operatives.

The Making Department

Although the making department in a turn-shoe factory was second in size to the stitching, it was the largest department employing exclusively men. Of the 294 operatives in the making department of our typical shoe factory, 278 were makers (lasters and beaters-out), 7 were Goodyear stitchers, 7 were tack-pullers and trimmers, and 2 were cobblers.

The making (or lasting) department was the main construction room. Here the processed soles and the stitched uppers, together with the counters, toe pieces, etc., were fastened together to form the rough shoe. Except for tack pulling and Goodyear stitching, the work here was done entirely by hand. The makers used many different hand tools and had to be proficient in more types of manual techniques than any of the other shoe operatives. Careful measurement by eye—without instruments —was required. Thus, to be a good maker required considerable experience. Competent lasting is essential to the production of a good shoe; no type of cobbling can correct fundamental errors of the lasters.

The makers performed the hardest manual labor of any operatives in the factory. They are known as the "blacksmiths" of the trade. They usually worked in teams of two: a laster and a beater-out. They divided the work between them and were paid as a team, half the earnings going to each man. There were instances where one individual performed both tasks, and in some factories three-man teams have been tried. These were not satisfactory, however, for it proved too difficult to discover and match three men of approximately equal speed and ability.

The first laster in the turn process dampened the sole for flexibility and tacked it to the bottom of the last with the outer part of the sole next to the last. He then inserted a counter between the lining and the outer leather of the heel of the upper to give strength to the finished shoe. When specifications called for it, he put a cement composition between the lining and the upper of the toe, and sometimes added a toe piece as well. These counters and toe pieces were also dampened before they were used. The laster used pincers to pull the upper, with the attached lining, inside out over the last and tacked the bottom of the upper and the lining to the sole around the toe and heel. This was entirely a hand process, and eye measurement alone was used.

The shoe, inside out over the last, with the upper temporarily tacked to the sole and tacked to the last at the top, was then passed to a worker at a Goodyear turn-stitching machine. The Goodyear stitcher fed the shoe to the needle of the machine, sewing the upper to the sole through the channel previously cut by operatives in the sole-leather department. This was as critical a job in shoemaking as lasting, for the quality of the stitching is important to the fit and capacity for wear of the shoe. Sewing was started on the shank and continued around the heel to the toe. One Goodyear stitcher could handle the output of twenty lasters. When the shoe was sewn, it was passed to the tack puller and trimmer. His machine removed the tacks from the sole and trimmed the rough edges of the upper.

The shoe was then returned to the first laster who pulled out certain tacks along the upper part of the last and then passed the shoe to the second laster or beater-out. The latter wore a thickly-padded apron since part of his work required body pressure. The beater-out removed the last and turned the shoe right side out. In the turning process he used a heavy metal spike which was attached solidly to his bench. He placed the heel of the sole on top of this spike and turned the heel of the upper right side out by pulling down on its outer edges. By gradually moving the spike forward on the sole and exerting pressure on the shoe itself, the beater-out was able to pull the entire shoe right side out. Before relasting, the beater-out inserted a filler in the forepart and a shank piece which had been coated with cement. The shank was fastened in place with two small nails. Then he relasted the shoe on its proper last.

After the shoe was relasted, the beater-out pounded and rolled the sole to smooth it, and with a light hammer pounded the upper to make it conform to the sole. He used a hot, pointed iron to crease the edges of the shoe between the sole and the upper and to shape it to the last. Next, he placed the shoe upside down on a bench form and, by further rolling and pounding, smoothed the sole into its proper shape. With the shoe still on the last, it was now ready to go to the heeling department. The last was not removed until the shoe was in the finishing department.

When the lasting was done by a team, the laster and the beater-out worked at the same bench, each with his own set of tools.[4] The makers were the only technological workers in a turn-shoe factory who ordinarily worked in teams and divided their combined earnings. The methods of pairing men into teams revealed interesting features of the social organization of the factory, particularly of the relation of operatives to their foremen. It was to the advantage of both the factory and the operatives to team men of approximately equal ability and speed.

Foremen were unanimous in saying that they completely controlled the pairing of makers, choosing their teams from their knowledge of the comparative ability and speed of the various men. To the operatives, however, the choosing of a team-mate seemed to be an informal process in which workers exercised their own choice. One man said: "You work with whoever you

---

4. These sets were said to be worth approximately twenty dollars apiece. Although the worker did not take them from his bench every night, they remained his property. This was the only department of a turn-shoe factory in which the ownership of tools represented a sizable investment by the operative; in other departments a worker rarely owned more than his own knife and such other small tools as might be needed in his particular process.

want. You get somebody you like to work with and start to make up a team. The foreman doesn't care how they work together." Another operative declared: "Sometimes, if one man doesn't have a partner, he'll say to the foreman, 'See here, I want a beater-out, somebody that's good,' and the foreman will say to a beater-out, 'Well, we have a maker here who wants a beater-out; if you can work with him, all right.' That's the way lots of us team up." These reports by workmen indicate that the foreman often picked the teams. Presumably if two men he teamed didn't work well together he would separate them or permit them to split up; but he would choose their new team-mates.

We mentioned earlier that one factory tried three-man teams of makers, with results that throw further light on the social structure of the factory. It proved very difficult to match men in threes who would work smoothly together. Often one of the three would delay the other two, and the workmen expressed a great deal of dissatisfaction with the arrangement. According to an official of the company, the management did not care whether two- or three-man teams were used as long as the men were satisfied. The company gave permission for the men to work either way they chose. But the operatives hesitated to return to the system of two-man teams, partly because they thought the company really wanted the three-man system, partly because they were afraid the two-man teams would be slower and earn less proportionately. The workers were apparently doubly (and contradictorily) suspicious: first, they suspected they would incur managerial anger if they insisted on returning to the two-man teams; second, they suspected the management felt that two-man teams were advantageous to their interests, disadvantageous to the operatives, so they were not sure, when they were given the choice, whether they really wanted to go back to the system of two-man teams.

The Heeling Departments

The process of heeling followed the construction of the main body of the shoe in the making department and varied according to the type of heel to be attached. In some factories heeling was done in two different departments: the wood-heel department and the leather-heel department. We shall describe these two departments separately because of the differences in the two heeling processes.

In keeping with the old New England tradition of shoemaking, until comparatively recently the wood heel was standard for women's shoes. Indeed, some factories used only wood heels before the depression of the early 1930's. The time required in the wood-heeling operation is much shorter than that for the construction job in the making room. Therefore, a factory needed fewer wood-heelers than makers. Leather-heeling, being a machine process, demanded even fewer men. There were forty-five workers, all men, in the wood-heeling department of our Yankee City factory.

The processes in the wood-heel department were performed entirely by hand, each operative doing exactly the same type of work. The technique depended upon the use of hand tools, especially the knife. The knife was the property of the worker, who beveled and shaped it to suit his method of work.

The wood-heeler first lifted up the loose piece that had been skived off the heel of the sole in the sole-leather department. He placed the heel upon the sole in its proper position and marked the sole with a knife. Removing the heel, he then beveled the outer part of the sole around the base of the heel. He applied liquid cement to the top of the heel, fitted it firmly, and placed the entire shoe in a clamp.

The shoe, thus held in position, was hung on an overhead rack allowing the cement to set properly. After a few hours the clamp was taken off, the loose flap of the sole was cemented carefully along the shank and up over the front part of the heel, and the edges of the flap were trimmed. The completed shoe with the heel attached was then passed to the finishing department.

There were eighteen men in the leather-heeling department of our typical factory. Whereas wood-heeling was done entirely by hand, leather-heeling was done entirely by machinery. Leather heels were received by the factory in sections, crudely cut to the shape of the heel. Each section contained layers of leather and composition substances. To make the heel the required height, a number of these sections were put together and pressed in a large machine. As the heel was held in the same machine, the shoe was placed on it and nails were driven into the heel, firmly attaching it to the sole. The machine was entirely automatic.

The "top lifts," forming the leather bottom of the heel, were riveted on by means of the so-called slugging machine.

A shaping or trimming machine shaped the rough, uneven surface of the heel. The operative held the heel, now attached to the shoe, against a revolving surface in which were set a number of curved knives. The operation was dangerous, because the knives revolved at an exceedingly rapid rate, and any slip by the operative might result in serious cuts on fingers or hand. The shoe was then passed to another operative whose machine trimmed the forepart, or breast, of the heel to its final shape.

The outer part of the heel was crowned and scoured by various machines. In some of these operations, fluids such as lacquer were applied as a base for the final finish. Some of the heel-finishing machines had cutting wheels; others had revolving surfaces which gave the heel an outside finish. In the last process in this department (or the first process in the finishing department) the shoe was fixed firmly in the sole-leveling machine which pressed the sole to its final shape.

The Finishing or Bottoming Department

Fifty-one male operatives worked in the finishing department. When shoes reached this department they were completely constructed and ready for the final trimming and finishing of bottoms and edges of the soles. Most of the processes were machine operations, although here, as in all departments, the shoe remained in the hands of the worker during the processing.

In some factories the sole-leveling machine was placed in the finishing room to perform the first operation in this department. Next, the shoe went to the edge-trimmer, who held the edge of the sole to a machine which had two revolving cutting knives of different shapes: the smaller trimmed the edge of the sole around the shank and heel, the larger trimmed the sole around the forepart.

From the edge-trimmer, the shoe passed to the edge-setter who stained the edge of the sole. Then he pressed it against a small rapidly oscillating iron, heated by a gas flame, which was part of a large machine. The purpose of this process was to set, harden, and give a finish to the outer edge of the sole. The operators of both the edge-trimming and edge-setting machine had to take care not to allow the revolving parts to come in contact with the upper part of the shoes, since any slip of this kind would spoil the whole shoe.

The buffer, next to get the shoe, held the sole to a revolving surface which smoothed the bottom. From the buffer the shoe passed to the operator of the "naumkeag machine" for another smoothing or buffing operation. This machine had two revolving pads, made of composition rubber and inflated with air.

The next operation was performed by the breasting machine. The operative applied the breast, or forepart of the heel, to a thin revolving surface to give it a final finish. The shoe was then passed to the next operative, a brush-stainer, after which the sole was ready for its final finish on the polishing wheels.

After the lasting tacks were pulled from the top of the shoe, and the last itself removed, a machine operation drove five nails through the upper part of the back of the heel into the sole. A hole was drilled from the inside of the shoe through the heel, and a screw inserted. When these processes were completed, the shoe went to the packing department where the uppers were cleaned, and the finished shoes were packed for shipment.

The Packing Department

This was a comparatively large department, giving employment to 42 men and 113 women. In a large factory the packing department was usually divided into two parts, the cleaning room and the packing room. The workers in the cleaning room were concerned with the final treatment of the uppers. The first operation consisted of trimming the lining. It was performed by a Booth trimming machine, which had a small, rapidly revolving knife.

In many styles of shoes there were cutouts in the upper leather over which the linings extended. This part of the lining was cut out by hand. After being thoroughly cleaned, the lining of the shoe was pressed smooth by an electrically heated iron. Sock linings, or thin inner soles, were cemented in the shoe.

Treers then placed the shoe upon the arm of a machine. The

outer extremity of the arm was fitted with an adjustable form or last. With the shoe firmly fixed in position, the treer cleaned and crudely polished the shoe, and applied a pointed iron, properly shaping the uppers where they joined the sole. The shoe then passed to a counter-shaping machine.

In the singeing process, which followed counter-shaping, a gas flame burned off any loose threads without marring the shoe. The oilers then applied oil to the outer part, and the dressers gave the shoe its final outer polish. The shoe was then passed to the repairers, who corrected any minor defects appearing in the finished upper. The edges of the shoe were cleaned, and a stick was inserted which held the shoe in proper shape.

The shoe was now ready for packing, but before being boxed in the packing room it was carefully inspected. Any necessary cobbling or minor repairing was done here. Then, each pair was wrapped in tissue paper and placed in a box stamped according to its size and style. The boxes were packed in cartons containing thirty-six pairs each.

The speed-up so apparent in the other departments of the factory was not generally evident in the packing department, although there were some exceptions. Many of the jobs in this department were hand jobs in which a group of people worked together and conversed almost continuously. More individuals in this department were paid on a time basis than in all the rest of the production departments put together. It might be that if time rates were changed to piece rates (as has been proposed in some factories) working speed would be increased in this department. It would be very difficult to set piece rates here, however, because a given operative does various jobs at different times, and it appears to be impracticable to set specific jobs for specific individuals in many cases.

The Designing Department

A factory usually employed one or more designers who were directly under the authority of the manager. Their work consisted of designing shoe styles or of adapting patterns to the particular requirements of the factory. Ordinarily, the designer was trained in a technological school, and in his work he might have much freedom of choice in his actions.

According to specifications he drew and cut out cardboard patterns from which the cutting department constructed the basic conform of sample shoes. As a rule, there were certain operatives in each department who worked on sample shoes. In most departments these operatives were paid on a time instead of a piece-work basis.

The designer customarily worked in or near the cutting department, not only because the precise nature of his work required as much light as possible, but also because most of the problems of designing were related to the uppers of the shoe. The designer was primarily a tool user. He used a variety of drawing tools and equipment, and a knife in cutting the cardboard dies.

Not all of the designs were created in the shoe factory. Some of this work had been taken over by specialized pattern companies. These companies made it unnecessary for the factories to employ many designers, and have tended to limit the work of those who were so employed to the adaptation of designs to meet the demands on their particular factory.

The Supply and Maintenance Departments

The workers in the supply department were principally object-handlers, although their work was in part concerned with the recording of the materials and supplies as they were received and delivered. This work was similar to the general type of work found in any business enterprise where there is need for handling and keeping records of materials. However, a knowledge of the materials and the terms used in a shoe factory is a necessity to the workers in the supply department.

The workers in the maintenance department were concerned with the upkeep of the plant and equipment. They were all technological workers. Some were artisans, such as carpenters, electricians, and engineers; and some were common workers, such as sweepers and cleaners. They took no part in the manufacturing processes and cannot be classed as shoe operatives. Their training or practice fitted them for jobs in any other large building as well as for jobs in a shoe factory.

APPENDIX 3

FREQUENCY OF MEMBERSHIP OF THE WORKERS

GENERALLY speaking, the shoe workers participated in more of the activities of the community than did the rest of the people. We examined the amount of worker membership in all the principal types of institutions in Yankee City. This appendix presents the results of studies of associational, clique, church, and political groupings.

That shoe operatives averaged more memberships per capita in associations than did the total adults of the same social stratification is shown by Table 7. It may also be noted here that among both shoe operatives and the total adult population there is a steady decline in number of memberships held per capita from lower-middle to lower-lower.

TABLE 7
Association Memberships Per Capita

	LM	UL	LL
Shoe operatives	1.67	1.29	.89
Total adult	.82	.55	.28

Among the shoe operatives who belonged to associations, 63.5 per cent belonged to one association, 25.5 per cent to two associations, 7.9 per cent to three, 1.8 per cent to four, and 1.3 per cent to from five to eight. The comparable figures for all members of the three lower classes do not vary materially from the shoe-operative figures in this instance. Among the individuals in the lower-middle and upper-lower classes a greater percentage of female than male, both among the shoe operatives and the total society, held memberships in one association only. In the lower-lower class a slightly higher percentage of the males held memberships in one association only in the total society than among the lower-lower shoe operatives, although there was a higher percentage of females who held memberships in one association among the latter. There was a tendency in all three classes, both among the shoe operatives and the total society, for more males than females to hold memberships in more than one association. Both among the shoe operatives and the members of the three lower classes of the total society the greatest percentage holding one membership only was in the lower-lower, the next highest in the upper-lower, and the lowest percentage in the lower-middle.

In the case of cliques, as in that of associations, shoe operatives held more memberships per capita than did the total adult population of the same social stratification. This is shown in Table 8. This table, also similar to that for associational memberships, shows that number of memberships per capita tended to decrease with descent in the social scale.

TABLE 8

Clique Memberships Per Capita

	LM	UL	LL
Shoe operatives	1.68	1.68	.91
Total adults	1.09	.91	.48

The most significant comparisons between the religious participation of the shoe operatives and that of the total population of the three lower classes came to light when we compared the professions of faith of the two groups. Four religious faiths are commonly recognized in Yankee City: Protestant, Catholic, Greek Orthodox, and Jewish. We found a large proportion of individuals who professed no religious faith both among operatives and in the total population of the three lower classes: 69.9 per cent of the shoe operatives and 58.5 per cent of all members of the three lower classes, specifically. This does not mean, of course, that such individuals never went to church. Moreover, many of those who professed no religious faith would have acted differently had our interviewers asked boldly whether they called themselves Protestants or Catholics. We did not force such a question in the belief that wholly free statements would give a more accurate picture of individual religious participation.

From the information obtained in this way we constructed Table 9, comparing the professions of faith of shoe operatives

TABLE 9

Professions of Religious Faith by Percentage

	Percent of Total Individuals Professing No Religious Faith	Distribution of Those Who Professed Religious Faith[1]			
		Catholic	Protestant	Greek Orthodox	Miscellaneous
Shoe operatives LM............	48.5	48.0	48.0	2.0	2.0
UL............	62.7	60.7	29.9	6.8	2.6
LL............	82.3	33.3	24.8	36.8	—
Total..........	69.9	50.9	33.8	13.5	1.8
Total Adult Y. C. LM............	50.3	32.2	60.3	0.3	7.2
UL............	54.9	58.3	33.3	2.1	6.3
LL............	72.1	44.4	43.6	9.4	2.6
Total..........	58.5	45.5	45.8	2.8	5.9

1. Since so few Jews held technological jobs in the shoe factories, the Jewish faith is not included here.

by social classes with those of the general population of the three lower classes in Yankee City. This table shows two points of interest: first, both among shoe operatives and in the total population of the three lower classes, profession of religious faith decreased progressively with descent in the social scale; and, second, the acceleration of this decrease was much greater among shoe operatives than in the total population. It was the relative lack of participation of the two lower classes of shoe [religion] (margin annotation)

operatives that caused the non-participation of all shoe operatives to be over ten per cent higher than that of the total population in the three lower classes.

Comparing the tabulations for those who professed religious faith throws further light on the constitution of the shoemaking group as compared with the total population of the three lower classes. There was a considerably higher proportion of Catholics among lower-middle shoe operators than in the lower-middle class generally. This is accounted for by the large number of Irish-Catholic women who worked in the stitching department, as described in an earlier chapter. The majority of the French shoe operatives were also Catholics.

The only significant difference between upper-lower shoe operatives and the upper-lowers in the total society in profession of religious faith was the higher proportion of Greek Orthodox among the shoe operatives. The figures for Greek Orthodox affiliation, it will be observed, show a persistent pattern throughout the column: in each class there was a higher proportion of Greek Orthodox among shoe operatives than among the total population of that class.

The fact of the recency of arrival of the Greeks accounts for the tremendous jump in the proportion of Greek Orthodox affiliations among lower-lower shoe operatives: as recent arrivals, they were predominantly lower-lower in social stratification; having employment largely as shoe operatives, they affected the distribution of professions of religious faith among lower-lower shoe operatives to a much greater extent than they did among the total lower-lower population. In the lower-lower figures, it is also to be noted that, whereas in the total population of this class the proportion of Catholics and Protestants was almost equal, among shoe operatives there were considerably more Catholics than Protestants.

Table 10 compares the percentage of voters among the shoe operatives with the total population of the three lower classes. It indicates a somewhat higher percentage of political participation among lower-middle and upper-lower shoe operatives than among the total population of these social classes. But a smaller percentage of shoe operatives than of total adults were voters in the lower-lower class; and, since members of this class formed such a large proportion of shoe operatives, the effect

TABLE 10

Political Participation by Social Classes

	Percentage of Registered Voters
Shoe operatives	
LM	58.5
UL	56.7
LL	28.7
Total	44.4
Total adult Y. C.	
LM	55.6
UL	50.8
LL	31.0
Total	47.2

of their non-participation was to bring the percentage of voters among all shoe operatives below the percentage for the total membership of the three lower social strata. It is probable that fewer of the lower-lower shoe operatives had taken out their citizenship papers than was the case in the general population.

APPENDIX 4

THE RANGE OF FAMILY MEMBERSHIP

IN considering the family memberships we use the concept of the family of procreation, which includes husband and wife and unmarried children living together as a unit under one roof.[1] In this analysis, in contrast to those of association and clique memberships, the number of family memberships held will coincide with the number of individuals.

Table 11 shows that, of the ninety-four shoe operatives of the lower-middle class, eighty individuals (85.1 per cent of the total) belonged to families all of whose members were lower-middle. This compares closely with 88.7 per cent for all lower-middle adults. Of the upper-lower shoe operatives, 86.3 per cent belonged to families all of whose members were upper-lower. This compares with 83.9 per cent for all upper-lower adults. So also, 96.4 per cent of the lower-lower shoe operatives belonged to families all of whose members were lower-lower, compared with 95.6 per cent for total lower-lower adults. These percentages do not vary sufficiently between shoe operatives and comparable total adult groups to be of much significance, but variations from single-class families require analysis. These variations are included in Table 11 which shows maximum range of social stratification of the families to which shoe operatives belong compared with similar figures for the total adult population of the three lower classes.

In Table 11 it will be observed that the percentage of lower-middle individuals who belonged to families that also had members above the lower-middle class was about the same for the shoe operatives as for the total adult lower-middle group. But a considerably higher percentage of lower-middle shoe operatives than of total adult lower-middle belonged to families which contained members who stratified below lower-middle. It is to be noted, moreover, that all variations from the single-class family for lower-middle shoe operatives extended only to

1. For a discussion of this concept see Volume I, pp. 28–35.

adjacent social classes, whereas in the total adult lower-middle class, there were families with members extending to upper-upper and other families which extended to lower-lower.

Similarly, the upper-lower shoe-operative families extended only to lower-middle, while in the total adult population some

TABLE 11

Total Number of Family Members by Social Stratification

		UU	LU	UM	LM	UL	LL	+	O	−
Shoe operatives (94)	LM	0	0	3	80	11	0	3	80	11
	%'s	0	0	3.2	85.1	16.7	0	3.2	85.1	11.7
(314)	UL	0	0	0	28	271	15	28	271	15
	%'s	0	0	0	8.9	86.3	4.8	8.9	86.3	4.8
(328)	LL	0	0	1	1	10	316	12	316	0
	%'s	0	0	0.3	0.3	3.0	96.4	3.6	96.4	0
Adult membership (3886)	LM	2	9	104	3456	299	30	115	3456	329
	%'s	0.1	0.2	2.6	88.7	7.7	0.7	2.9	88.7	8.4
(4235)	UL	0	0	24	507	3565	163	531	3565	163
	%'s	0	0	0.5	11.9	83.9	3.7	12.4	83.9	3.7
(2940)	LL	0	0	4	30	98	2808	132	28.08	0
	%'s	0	0	0.1	1.0	3.3	95.6	4.4	95.6	0

upper-lower families had members who extended into upper-middle, and so exhibited a wider range. Moreover, the total figures for these two upper-lower groups showed that a greater percentage of upper-lower shoe operatives than of total upper-lower adults belonged to families which contained members who stratified below upper-lower. This shows that upper-lower shoe-operative families tended, to a greater extent than did the upper-lower families in the total population, to maintain the one-class family; where shoe-operative families departed from this norm they tended to orient to lower-lower to a greater extent than did upper-lower families generally.

The family affiliations of lower-lower shoe operatives varied only slightly from those of the total adult lower-lower population. Even here, however, what little indication there is suggests a tendency for a smaller percentage of lower-lower shoe operatives than of lower-lower members of the total adult population to belong to families which contained members above the lower-lower class.

Table 12 gives the breakdown for the more numerous ethnic groups, showing the percentage of memberships in families which varied in range from the usual one-class family and comparing the figures for shoe operatives with the figures for the total Yankee City population of all social strata and including the sub-adults. The figures are percentages of total family memberships in families which include members of more than one social class.

TABLE 12

Percentage of Memberships in Families Which Varied in Range from the One-Class Norm, Showing Variations by Ethnic Groups between Shoe Operatives and Total Yankee City Population

	Irish	Italian	French	Greek	Native Yankee
Shoe operatives	20.26	9.68	9.56	7.32	4.83
Total Yankee City (including the immature and all social strata)	17.13	9.71	6.84	6.88	7.07

This table shows that the Irish were by far the most socially mobile ethnic group in Yankee City, both among shoe operatives and in general, while the oldest resident group (the native Yankees) and the newest ethnics (the Greeks) were the least mobile. It also shows that, excepting Italians and Yankees, families of shoe operatives were more mobile than families in the general population belonging to the same ethnic groups.

APPENDIX 5

ETHNIC WORKERS' MEMBERSHIPS

TABLE 13 brings out several significant facts about ethnic workers' memberships. First, it will be noted that in every case shoe operatives held more memberships per capita than did the comparable group in the total population. Second, with a few exceptions, lower-middle individuals, both among shoe operatives and in the total society, averaged more memberships per capita than upper-lower, who in turn exceeded lower-lower. Third, the Irish, both among the shoe operatives and in the society at large, held more memberships per capita than other ethnic groups of the same social strata. The French were second in per capita number of memberships both among the shoe operatives and in the total population of the three lower classes. The native Yankees, Italians, and Greeks were roughly tied for third place in per capita memberships in both groups.

The fact that per-capita memberships among the shoe operatives exceeded those among comparable groups in the total population would seem to indicate that shoe operatives participated more actively in the social life of Yankee City than did average members of the three lower classes. But the class-types of memberships held by the shoe operatives, as we have shown in the tables giving extreme range of social participation, actually reflect the fact that even outside the factory they remained fairly distinct from the general population of the classes to which they belonged. The differential participation by ethnic groups, since it was consistent among both shoe operatives and the general population of the three lower classes, is to be interpreted as reflecting a differential in the traditional training of individuals in the various ethnic groups regarding the value of memberships in organizations.

Thus we see that being a worker in a shoe factory modified the operative's social life in the community at large both qualitatively and quantitatively. He behaved differently from other individuals in the community of comparable place in the social stratification. The effects of the shoe factories on the social lives of operatives demonstrate that these factories must be considered as social institutions, not as merely economic institutions. They demonstrate, too, that it does not suffice to discuss

TABLE 13[1]

Memberships Per Individual, Shoe Operatives Compared with Total Adult of Three Lower Classes, Shown by Social Class, Ethnicity and Totals

	Irish	French	Italian	Greek	Total Ethnic	Native	Grand Total
Shoe operatives							
LM	6.04	5.13	6.33	4.50	5.70	5.20	5.43
UL	5.56	5.19	4.82	4.57	5.24	4.11	4.90
LL	4.41	3.02	2.65	3.56	3.46	2.94	3.26
Total, Three Classes	5.48	4.24	3.77	3.84	4.51	3.77	4.24
Total adult Y. C.							
LM	4.85	4.04	5.43	3.04	4.79	3.91	4.23
UL	4.27	3.52	3.13	3.26	3.93	3.18	3.67
LL	2.85	2.32	2.08	2.71	2.60	2.34	2.50
Total, Three Classes	4.24	3.07	3.00	2.93	3.68	3.29	3.50

1. This table includes only the major ethnic groups among the shoe operatives.

only the formal, "official" relations between the community and management if we would fully comprehend the effect of a factory on a community. The informal impingement of the factory, through its technological workers, on the social behavior and attitudes of the community show that much more searching analyses must be made to show the roles of factories in the social structure of communities such as Yankee City. The informal integration of the shoe factories into the social structure of Yankee City has broad *social* implications which have not been much considered in previous studies of American industry.

APPENDIX 6

SOCIAL CHARACTERISTICS OF THE SHOE WORKERS

MOST of the salient social characteristics of shoe operatives are summarized in Table 14. This appendix, describing the operatives as a whole, will consist primarily of a discussion of particularly relevant points in this table. In addition, however, some significant factors which do not appear therein will be discussed in the text.

The proportion of our sample who were males (57.5 per cent) was slightly above the average for the shoe industry as a whole, in which 55.0 per cent of the operatives were men. This is doubtless due largely to the fact that they were in factories which used the turn process, entailing more hand work than the newer methods, in which older people and men are usually preferred.

The median age of both men and women was likewise higher in this factory than in the industry as a whole. Again, this was probably due in large part to the use of the turn process. Some of the techniques require considerable practice; hence the workers at these jobs were not young people. More important, perhaps, was the waning popularity of the turn method. Since there was already an oversupply of workers who knew the turn process, younger workers tended to learn other manufacturing techniques.

The median age for both sexes is presented in the table. Our detailed tabulations, not presented here, show that women operatives were more numerous than men in all age brackets below forty years, and that men were more numerous in all age brackets above forty years. This is doubtless accounted for by the tendency of women to leave the factory when they marry. Among specific ethnic groups, the low median ages for both the French and Polish are explained by the fact that they contained large proportions of women operatives and also, in all probability, by the tendency of these operatives toward early mar-

TABLE 14[1]
The Social Characteristics of Shoe Workers

		Number of Workers		Social Stratification								Place of Residence						Median Age
				LM		UL		LL		Unknown		In Yankee City		Outside Yankee City		Unknown		
		Men	Women	Men	Women	Men	Women	Men	Women	Men	Women	Men	Women	Men	Women	Men	Women	
Yankee	No.	286	148	20	18	60	42	93	38	113	50	112	101	172	45	2	2	41.1
	% of Total	29.1	15.0	2.8	2.5	8.3	5.8	12.8	5.3			39.2	68.2	60.1	30.4	.7	1.4	
Combined ethnics	No.	280	271	15	20	97	116	103	101	65	34	207	238	68	29	5	4	38.5
	% of Total	28.3	27.5	2.1	2.8	13.4	16.0	14.2	14.0			73.9	87.8	24.3	10.7	1.8	1.5	
French	No.	89	95	4	5	32	36	23	29	30	25	47	73	39	21	3	1	29.9
	% of Total	9.0	9.7	.6	.7	4.4	5.0	3.2	4.0			52.8	76.8	43.8	22.1	3.4	1.1	
Irish	No.	49	92	6	11	28	70	10	10		1	43	86	6	6			38.7
	% of Total	5.0	9.4	.8	1.5	3.9	9.6	1.4	1.4			87.8	93.5	12.2	12.2			
Greek	No.	74	31	1	0	22	4	39	25	12	2	59	30	14		1	1	42.3
	% of Total	7.5	3.1	.1		3.0	.6	5.3	3.5			79.7	96.8	18.9		1.4	3.2	
Italian	No.	32	7	1	2	8	2	12	3	11		23	7	8		1		42.2
	% of Total	3.2	.7	.1	.3	1.1	.3	1.7	.4			71.9	100.0	25.0		3.1		
Polish	No.	8	31				2	6	27	2	2	7	29	1	1		1	24.4
	% of Total	.8	3.1				.3	.8	8.7			87.5	93.6	12.5	3.2		3.2	
Other	No.	28	15	3	2	7	2	13	7	5	4	28	13		1			43.4
	% of Total	2.8	1.5	.4	.3	1.0	.3	1.8	1.0			100.0	86.8		6.6			
Totals	No.	566	419	35	38	157	158	196	139	178	84	319	339	240	74	7	6	All Men 43.0 / All Women 35.4
	% of Total	57.5	42.5	4.8	5.3	21.7	21.9	27.1	19.2			56.4	80.9	42.4	17.7	1.2	1.4	
	No.	985		73		315		335		262		658		314		13		Total 39.6
	% of Total	100.0		10.1		43.6		46.3				66.8		31.9		1.3		—

1. These figures are derived from two sources: actual factory records and other regions of the Yankee City...

riage. In the case of the Polish group there is the further factor to be noted: they were comparative newcomers to Yankee City; most of the Polish operatives are second generation, hence of necessity young. The Irish, on the other hand, have been in Yankee City longer than any other ethnic group, long enough for shoemaking to become a traditional occupation of many, particularly the women. Hence a number of Irish women operatives were middle-aged. The same was true of the native Yankee women operatives.

Table 14 indicates that only 44.1 per cent of the shoe operatives were native Yankees; the other 55.9 per cent were ethnics. The proportion of women ethnics to total women (64.7 per cent of all women) was much greater than the comparable proportion in the case of men (49.5 per cent of all men).

The large number of Irish and French in the shoe trade is a reflection of the fact that they came to Yankee City before other ethnic groups, when shoe manufacturing was even more important than it is now as a means of gaining a livelihood. Members of these two ethnic groups, from this circumstance, have almost as strong a tradition of making shoes by the turn process as have the native Yankees. Among both the French and the Irish, however, and especially in the latter group, women form a much higher proportion of the operatives than among the native group. This indicates two things: first, that technological work done by women is more characteristic of ethnic peoples than of natives; second, that there still exists a stronger tradition of handicraft among natives than among ethnics. Since women are very largely confined to machine jobs, we may conclude that there is a different sex-wise selection between native and ethnic operatives.

The social stratification of individuals in Yankee City society has been shown in other volumes of this series to be a very important factor in predicting the type of social behavior and the roles of individuals in the life of the community. To this general rule the shoe operatives are no exception (see Chart XII). The social stratification of the shoe operative is of particular interest because this group shows important deviations in its social behavior from other Yankee City residents of comparable social class. In other words, we must take into account special modifying factors which affect the social behavior of shoe operatives and their position in the social hierarchy (see p. 159). An examination of the social stratification of male and female shoe workers shows that women operatives stratify higher in the social scale than men (see Chart XIII). A probable partial explanation for this fact is that the opportunities for women to get jobs in Yankee City were more limited than were opportunities for men. It may be that women of higher social strata—if they had to work—were given preferential treatment by employers in the shoe factory.

The high proportion of ethnics in technological jobs in the shoe factory goes a long way to account for the low overall social status of shoe operatives, since the newer ethnic groups

CHART XII

Social Stratification of the Shoe Operatives as Compared with Total Yankee City Community

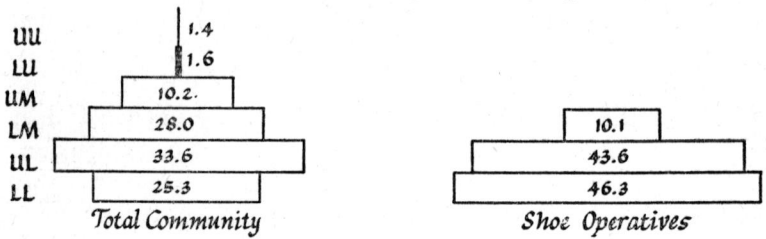

belong predominantly to the lower-lower or the upper-lower classes as is shown in Table 14. Only one Greek and three Italians in the shoe factory were rated as lower-middle, although Yankees made up 52.1 per cent of the shoe operatives in this class. The latter were about evenly divided as to sex, although the number of native men in the factory as a whole was much greater than the number of native women. The rest of the lower-middle shoe operatives were predominantly Irish and French who are "foreigners" only to those at the very top of the Yankee City social hierarchy.

In the upper-lower stratum, the proportion of native Yankees drops to slightly less than a third. There were almost as many Irish as natives in this social class in the shoe factory. The majority of the Irish women in the factory belonged to this social class. The French were also numerous in this class, and the number of Greek males was increasing; very few Italians, Poles, or "other" ethnics had yet made the grade. Noteworthy is the fact that of the 213 ethnics who rated as upper-lower, 116 were women, a higher proportion than would be expected from the proportion of the sexes in the total number of the workers.

Yankees made up approximately one-fifth of the factory workers who stratify as lower-lower. Of these, the men outnumbered the women by nearly three to one, although the sexual ratio of all natives in the factory was about two to one. In other words, native men in the factory stratify lower than do native women. In this class we find a considerable number of

CHART XIII

Percentage Figures of Social Stratification of Shoe Operatives by Sex

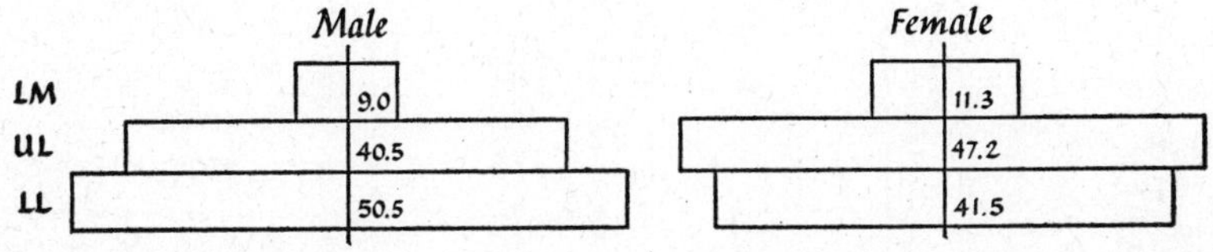

	Male	Female
LM	9.0	11.3
UL	40.5	47.2
LL	50.5	41.5

French about evenly divided in sex, comparatively few Irish, and the bulk of the Greek, Italian, Polish, and "other" ethnics.

In examining the participation of shoe operatives in the community life of Yankee City, it is essential to note how many operatives live in the community and how many come to work from outside. Obviously, the social and political loyalties of non-residents of Yankee City center in their home communities. Yankee City, therefore, has less control over such individuals, and the larger their number in the factory the greater is the social distance between factory and community. Our analysis showed a high proportion of the shoe factory operatives, 32.3 per cent, were non-residents of Yankee City. Of these 314 individuals, 240 (76.4 per cent) were men and only 74 were women. Two-hundred and seventeen non-residents were native Yankees, ninety-seven were ethnics.

The social characteristics of shoe operatives as a whole may be summarized as follows: slightly more than half of the oper-

TABLE 15

Social Characteristics of the Technological Workers in Different Departments

	Sole Leather		Cutting		Stitching		Assembling		Making		Wood Heeling		Leather Heeling		Finishing		Packing		Totals	
	No.	% of Total	No.	% of Total	No.	% of Total	No.	% of Total	No.	% of Total	No.	% of Total	No.	% of Total	No.	% of Total	No.	% of Total	No.	% of Total
NUMBER OF WORKERS																				
Men	14	100	80	100	7	2.2	15	100	294	100	45	100	18	100	51	100	42	27.1	566	57.5
Women	—	—	—	—	306	97.8	—	—	—	—	—	—	—	—	—	—	113	72.9	419	42.5
SOCIAL STRATIFICATION																				
LM	—	—	9	14.5	27	11.0	1	10.0	8	4.1	5	20.0	3	37.2	6	16.7	14	10.9	73	10.1
UL	7	70.0	38	61.3	115	46.9	3	30.0	60	30.6	10	40.0	4	36.4	20	55.5	58	45.3	315	43.6
LL	3	30.0	15	24.2	103	42.1	6	60.0	128	65.3	10	40.0	4	36.4	10	27.8	56	43.8	335	46.3
Unknown	4	—	18	—	68	—	5	—	98	—	20	—	7	—	15	—	27	—	262	—
ETHNICITY																				
Native Yankee Men / Women	5 / —	35.7	43 / —	53.8	1 / 106	34.2	9 / —	60.0	148 / —	50.3	33 / —	73.3	8 / —	44.5	32 / —	62.7	7 / 42	31.6	286 / 148	44.1
Combined Ethnics Men / Women	9 / —	64.3	37 / —	46.2	6 / 200	65.8	6 / 0	40.0	146 / —	49.7	12 / —	26.7	10 / —	55.5	19 / —	37.3	35 / 71	68.4	280 / 271	55.9
French Men / Women	4 / —	28.6	18 / —	22.5	— / 76	24.3	1 / 0	6.7	45 / —	15.3	6 / —	13.4	2 / —	11.1	11 / —	21.6	2 / 19	13.5	89 / 95	18.7

TABLE 15 (Continued)

	Sole Leather		Cutting		Stitching		Assembling		Making		Wood Heeling		Leather Heeling		Finishing		Packing		Totals	
	No.	% of Total	No.	% of Total	No.	% of Total	No.	% of Total	No.	% of Total	No.	% of Total	No.	% of Total	No.	% of Total	No.	% of Total	No.	% of Total
Irish Men	3	} 21.5	13	} 16.2	1	} 18.8	3	} 19.9	9	} 3.1	3	} 6.7	4	} 22.2	8	} 15.7	5	} 25.2	49	} 14.3
Women	—		—		58		0		—		—		—		—		34		92	
Greek Men	1	} 7.1	—	} 0	—	} 8.0	1	} 6.7	60	} 20.4	—	—	2	} 11.1	—	—	10	} 10.3	74	} 10.6
Women	—		—		25		0		—		—		—		—		6		31	
Italian Men	—	—	1	} 1.3	1	} 2.2	—	—	24	} 8.2	1	} 2.2	2	} 11.1	—	—	3	} 2.6	32	} 4.0
Women	—		—		6		—		—		—		—		—		1		7	
Polish Men	—	—	1	} 1.3	—	} 8.0	—	—	3	} 1.0	2	} 4.4	—	—	—	—	2	} 5.2	8	} 4.0
Women	—		—		25		—		—		—		—		—		6		31	
Other Men	1	} 7.1	4	} 4.9	4	} 4.5	1	} 6.7	5	} 1.7	—	—	—	—	—	—	13	} 11.6	28	} 4.3
Women	—		—		10		0		0		—		—		—		5		15	

PLACE OF RESIDENCE

	No.	% of Total	No.	% of Total	No.	% of Total	No.	% of Total	No.	% of Total	No.	% of Total	No.	% of Total	No.	% of Total	No.	% of Total	No.	% of Total
In Yankee City	9	64.3	62	80.5	252	81.6	8	53.3	145	49.8	17	37.8	9	50.00	30	60.0	126	82.3	658	67.7
Outside Yankee City	5	35.7	15	19.5	57	18.4	7	46.7	146	50.2	28	62.2	9	50.00	20	40.0	27	17.7	314	32.3
Unknown	—	—	3	—	4	—	—	—	3	—	—	—	—	—	1	—	2	—	13	—

MEDIAN AGE

Men	32.5		40.4		36.3		28.5		43.7		37.7		49.6		33.8		44.1		43.0	
Women	—		—		35.0		—		—		—		—		—		36.0		35.4	

Appendices 233

atives were males whose average age was higher by a few years than that of the women; almost exactly one-half of the men were ethnics whereas two-thirds of the women may be so classified; the women, however, were drawn in general from one social class above that of the men; and, finally, one-third of the workers—a predominantly male group—were non-residents of Yankee City.

Table 15 tells the story of the similarities and diversities that existed among the workers of the several departments. It is a complex table but easy to study if the reader will notice that it is divided into five main headings. At the top of the page is the number of workers followed by social stratification, ethnicity, place of residence, and ending with median age at the bottom of the page.

The grand totals and percentages are in the column at the extreme left; the data on the departments are listed in the column to the right of it.

Analysis of the social characteristics of the workers in various departments brought out certain important facts about them that were not disclosed by the analysis of all workers considered together; there were important differences among the various departments. In our sample, three departments—stitching, making, and packing—accounted for two-thirds of all workers. These departments were also high in ethnicity: two-fifths of the stitchers and of the packers were ethnic (all the women employed were in these departments); half of the makers and cutters were ethnic (see Table 15).

In social stratification, cutting ranked highest of the large departments, stitching and packing tied for second, and making was a poor third. The position of cutting is undoubtedly correlated with the shoemaking tradition that cutters are the aristocrats of the trade. The positions of stitching and packing are primarily due to the large numbers of women in these departments—women in the shoe factory stratify higher than men, on the whole. Making has such low social stratification, partly because so many of its ethnic members belong to the newer ethnic groups (there are almost no Irish in this department, for example), partly because the mark of a maker—badly calloused hands—both selects lower-lower individuals for makers' jobs and prevents makers from rising in the social scale.

Of interest, too, is the fact that the three large departments that rank highest in social stratification—cutting, stitching, and packing—had the largest proportion of Yankee City residents of all departments in the factory, while the making department, such a poor fourth in social stratification, is near the bottom of the list in proportion of resident operatives.

INDEX

ABC Company, 141, 155–156; and Jones and Jackson Company, 147; and Shoe Workers' Protective Union, 149; in strike, 40, 42, 48, 49, 50; and Weatherby and Pierce Company, 140. *See* Managers, absentee; Owners, absentee

Absentee capital, 27, 152–153, 154–155, 187. *See* Owners, absentee; Managers, absentee

Age-grade structure, 87–89. *See* Craftsmanship

Age groups, as wage factor, 103

American Federation of Labor, 123, 127–128

Antagonisms, inter-class, 18; of workers, 14, 16–17, 19–21, 24, 27

Apprenticeship, function of, 57, 79, 87. *See* Craftsmanship

Arbitration, sanctions of, 194. *See* State Labor Board of Arbitration and Conciliation

Armenians, and management, 142; in the strike, 5. *See* Ethnic workers

Associations, 166; community, in the strike, 35; employers', 122–123; horizontal, 123–124; horizontal and vertical extensions of, 196; manufacturers', 108, 109, 121, 122, 124, 131, 132; memberships of shoe workers and of the three lower classes, 161, 166–167, 218; trade, 187. *See* Union

Banks, in the depression, 10

Bargaining power of workers, through collective action, 177; and social stratification, 103; and supply and demand for labor, 98–100, 102; and unified demands, 28; and worker solidarity, 93–95, 100

Bronstein's shoe factory, 33, 39, 40, 44–45, 47, 48, 140, 147, 152. *See* Managers, absentee; Owners, absentee

Business, controlled by state and federal governments, 187; hierarchies expanded, 193; internal organization, 187; and trade associations, 187. *See* Industry

Capitalism, in conflict with labor, 194; international, and cartels, 195; regulated, 187

Capitalists, factory owners, 27; in history of shoe production, 65; in history of Yankee City, 54

Catholic Church, in relief activities, 16; memberships of shoe workers and of three lower classes, 170–171, 218–220

Chamber of Commerce, in the depression, 8; in the strike, 35, 48; and shoe manufacturers, 22–23

Charitable organizations, local, in the depression, 8, 13, 14; and social status, 14–15. *See* Public welfare

Churches, in the depression, 10; relief activities, 14, 16; status hierarchy, 188; vertical and horizontal extensions, 196. *See* Catholic; Protestant

Citizens, committee in the strike, 50; in the depression, 10, 13; and relief agencies, 15; workers as, 160. *See* Townspeople

Cliques, 168; among ethnic workers, 94–95; memberships of shoe workers and of three lower classes, 161, 168–170; in shoe factory departments, 94–95, 103

Closed shop, as union objective, 40, 43

Collective bargaining, to control management, 70, 177

Committee of Industrial Physiology of Harvard University, ix

Commons, John R., 54, 60

Community, control of factories, 48, 49, 108, 109, 113, 119, 122, 152, 178–179; in the depression, 12, 13; and factory as social institution, 160; and industrial integration, 122; job evaluation, 90, 118; and local managers and owners, 118; status of workers, 159; in the strike, 6, 37–38; systems and social planning, 193

Community Welfare, in the depression, 13, 14

Conflict, between capital and labor, 194; institutionalized, 43, 109; interclass, 18; between management and unions, 122–123; between unions (CIO–AFL), 123; between workers and managers, 109, 131, 181, 193; among workers, 178

Congress of Industrial Organizations, 123, 127–128

Control, of business by government, 187; of family, 193. See Social control

Cost of living, and wage cuts, 24–25

Craft unions, 65, 125, 127, 128; and technological hierarchies, 176–177

Craftsmanship, 76, 87–88, 98, 119; and apprenticeship, 57, 79, 87; and mechanization, 61, 79, 83, 86, 89. See Master craftsmen; Skill

Crisis, social and industrial, 1, 6; economic, 15

Democratic ideology, 172, 189

Depression, and charity, 13–14; and class animosities, 24; emergency employment, 13; and loss of industry, 49, 63; and market stimulation, 26–27; and price reductions, 25; and relief, 13, 14; and social reorganization, 180; and social status, 19, 20–21; spread work, 8; and the strike, 4, 8–24; and style changes, 26–27; and upper classes, 9, 11, 12, 13, 18, 20–21; wage reductions, 23, 25, 30

Donham, Dean Wallace Brett, ix

Durkheim, Émile, 54

Economic change, and relations of power and prestige, 189; and social class, 181, 187–188

Economic determinants of the strike, 4, 6, 7, 21, 53

Economic history of Yankee City, 2, 54–56; apprenticeship system, 57; Clipper Ship era, 54; comb manufacturing, 57–58; cotton textiles, 58; decline of maritime trade, 56; in the depression, 10, 13; handicrafts, 56; machine methods, 57; manufacture of shoes, 57, 58; migration of factories, 62–63; simple folk economy, 54; social integration, 35; steam power, 57–58; welfare in the strike, 49

Economic organization of workers, and social change, 188

Economic system, international, 195–196; upper-class control, 22

Employer-employee relations, 71, 130; employment contract, 69; "free contract," 187; former, 138–139; leveling of technical jobs, 80; and social controls, 67, 68–69, 108. See Management; Workers

Employment, emergency, 13; part-time in the depression, 9; steady, and absentee owners, 121

Ethnic workers, children of, 17; cliques, 94–95; differential prestige, 96–97; favored by management, 97–98; "foreigners," 97, 101; power and prestige, 3; social characteristics, 227–235; social participation, 225–226; social solidarity, 92–93. See each ethnic group

Factory system, in Yankee City, 2; and extension outside, 108–133. See Shoe factory

Family, as institutional control, 193; of shoe workers and lower classes, 161, 222–224; systems and social planning, 193; and women in jobs, 18

Fears of workers, 14, 16–17, 19–20

Federal agencies, CCC and CWA, 14, 15; ERA, 15. See Government, federal

Foreigners, antagonism to, 94; as factory managers, 23, 157; as "newer ethnics," 97; and the strike, 5. See Ethnic workers

Foremen, 68, 71–72, 73, 83, 96, 174–175; and ethnic-worker solidarity, 95; in the strike, 32, 42; and women employees, 91; and worker relations, 66, 79, 86

"Free contract" in labor-management relations, 107

French Canadian workers, and family, 224; and management control, 97; memberships, 225–226; pay differentials, 98; prestige, 96, 119; social class, 3; social solidarity, 93, 95; in the strike, 5. *See* Ethnic workers

Ginsberg, M., 54
Government arbitrators, 6; of U. S. Department of Labor, 47. *See* State Labor Board of Arbitration and Conciliation
Government, placing of power, 193–194; planning for, 191
Government, federal, CCC and CWA, 14, 15; control of business, 187; destruction of power, 194; relief, 14, 16; as separate system, 194; Department of Labor, 47. *See* New Deal; President Roosevelt
Government, local, in the depression, 13, 14; relief agencies, 14–15. *See* Mayor; Police
Government, state, and control of business, 187. *See* State Board of Arbitration and Conciliation
Greek Orthodox shoe workers, 218–220
Greek workers, 4; in cliques, 95; and ethnic prejudices, 93; family, 224; and management, 142; memberships, 225–226; religious affiliation, 218–220; social class, 3; in the strike, 5. *See* Ethnic workers
Grievances of workers, regarding piece rates, 26–27, 28; spread work, 24, 25

Handicraft, 79, 81. *See* Craftsmanship
Harvard Graduate School of Business Administration, ix
Hierarchy, associational, 195, 196; corporate, 193; craft, 87–88, 175; ecclesiastical, 195, 196; economic and social, 181, 188, 195; of factory system, 108, 157, 197; of horizontal associations, 123; industrial, 123, 190; managerial, 113; manufacturing, 110, 121; political, 195, 196; social, 19, 65, 183, 194, 195; of status and skill, 66, 74, 76, 81, 125, 127, 175, 182, 189; of supervisory controls, 68–69, 110–111; of technological jobs, 67, 80–81, 110–111, 127–128, 176–177; of unions, 127–128, 177, 193
Hill Streeters, 3; attitude towards, 10; former factory owners and managers, 155. *See* Upper class
Hobhouse, L. T., 54
Horizontal extensions, 123, 132; of economic institutions, 195; through labor unions, 108, 109, 122, 132; through manufacturers' associations, 108, 109, 121, 124; of shoe factories, 121, 122, 178–179; of social order, 193
Human society, forces to control, 1; planning for, 191; in rational and logical world, 192; and species behavior, 192

Incomes of shoe operatives, 25. *See* Wages; Pay
Industrial activity, in the depression, 13; in the strike, 46, 48
Industrial society, and class relations, 159–160; integration of, 122; nature of, 6; and role of local community, 121
Industrial union, 2, 52, 125, 127, 128; as horizontal extension, 125; and technological hierarchy, 176–177; United Shoe and Leather Workers' Union, 125–127
Industry, and destruction of skill hierarchy, 189; hierarchy, 190; and human factor, 192; planning for, 191; and specialization, 189, 190; and technological advances, 190
Irish workers, and family, 224; and management control, 97, 141; memberships, 225–226; and pay differential, 98; prestige, 96, 119; social class, 3; social solidarity, 93; in the strike, 5. *See* Ethnic workers
Italian workers, in cliques, 95; and family, 224; memberships, 225–226. *See* Ethnic workers

Jews, as absentee managers, 10, 23, 42, 140–141, 147, 148; in social class, 3; in the strike, 5; as workers, 218
Job evaluation, by community, 90,

105–106, 175; by management, 67, 90, 105–106; and rates of pay, 66, 67, 90, 105–106; by workers, 66, 67, 90, 105–106. *See* Hierarchy; Skill

Job-making in the depression, 13

Jones and Jackson shoe factory, 32, 138; and ABC Company, 147, 154; and mechanization, 25; and piece rates, 27; in the strike, 43, 50, 145–147, 154; and workers, 41. *See* Managers, absentee

Knights of St. Crispin, 61, 88, 125

Labor, in conflict with capital, 194; division of, 64, 66, 68–69, 76, 80, 86, 109–110, 111, 190, 194; and industrial unionism, 125, 127, 177; and prestige of shoemaking, 119; and subordination of workers, 175; hand and machine, 60–61; organization, 61; as separate system, 194; sexual division of, 90; supply and demand and wages, 98–99. *See* Workers; Union

Land, John, as factory manager, 27, 32, 41, 140, 154; in the strike, 38–39, 45–47, 141, 143–144; and the workers, 142

Lower class, antagonism to upper class, 10–11, 17, 19–21; associational membership, 161, 166–168; associations in the strike, 35; clique membership, 161; in the depression, 9–10, 14–15, 16–17; family structure, 161; and federal relief, 14–15, 16–17; merchants in the strike, 36–37; neighborhoods, 19; police, 43; religious participation, 161, 170–171; Riverbrookers, 3; social relations, 159–160, 166, 177–178; and upper-class control, 22; workers, 9, 17, 18, 22, 33, 103, 167–168, 169–171, 217–221, 222–224, 225–226, 228, 230, 231, 232–233

Luntski shoe factory, 33, 52, 140, 148, 152. *See* Managers, absentee

McKay, Gordon, and leasing system, 62

Management, attacked in strike, 32; in conflict with workers, 1–2, 4, 6, 28, 29–30, 66–67, 109; control over workers, 31, 33, 63, 68–69, 78–79, 80, 81, 97–98, 131; and ethnic employees, 97–98; evaluation of jobs, 67, 83–84, 90; function, 71; and leasing machines, 62; and mechanization, 66, 80; and mediation in the strike, 51, 52; and National Boot and Shoe Manufacturers' Association, 47; and piece rates, 27, 102; profits, 27, 172–173; solidarity, 44; and spread work, 25–27; strategy during strike, 31–32, 33, 38–40, 42, 43; and strike breakers, 29; against union, 7, 29, 31, 122; -worker relations, 9, 25, 63, 86, 88, 108, 114, 117, 187, 190, 193

Managers, absentee, 5, 33, 112, 115–117, 140, 152; and cause of strike, 21; and external organization of factory, 112–113, 115–118; freedom from community control, 48, 49, 108, 109, 113, 115, 118–119, 152–153, 179; and internal organization of factory, 112–113, 115–118; Jewish, 10, 33; and piece rates, 27; profits, 27, 119, 179; and removal of factories, 36, 49; and steady employment, 121; in the strike, 38, 43; workers' distrust of, 9, 23–24, 27, 42, 47; -worker relations, 108, 110, 119, 179. *See* Luntski; Bronstein; Jones and Jackson; ABC Company

Managers, former, 5; Caleb Choate, 135–137, 140, 141, 142, 150–152, 155, 156; as collective representation, 134, 151, 156; and community control of industry, 152; in employer-employee relations, 138–139, 153; history of, 135–139; William Pierce, 135, 138–139, 140, 142, 152, 153, 156; in the strike, 134, 151; as symbols, 151; Godfrey Weatherby, 135, 136–138, 143, 152, 153, 156

Managers, local, 32–33, 112, 115–116, 140, 154; and cause of strike, 21; and community control, 118; and external organization of factory, 111–112, 115–118; internal organization of factory, 111–112, 115–118; and piece rates, 27; power, 193; profits, 27; in the strike, 4, 5; and worker security, 9, 23. *See* Weatherby and Pierce; Jones and Jackson

Index

Manufacturers' associations, 108, 109, 121, 122, 124, 131, 132
Market, and mechanization, 60; and wage cuts, 25, 27
Master craftsmen, 57, 74, 76, 81. *See* Craftsmanship; Skill
Mayo, Professor Elton, ix
Mayor, coöperation in the strike, 37–38, 39–40; in the strike, 34, 42, 47, 50, 52, 149. *See* Government, local
Mechanization, and blocked worker mobility, 79–80, 88; as cause of depression, 21, 25; and expanding market, 60; and industrial unions, 127, 176–177; and inter-worker relations, 77–78; and job leveling, 80–81, 175, 176, 190; and labor costs, 60–61, 106–107; and larger economic structures, 190; and operators, 76–80, 174, 190; and prestige of shoemaking, 119; and skill, 66, 76; and social structure of factory, 60, 111; and the strike, 21; union attitudes, 37; and worker-manager relations, 66, 78, 80; worker attitudes, 25, 77
Meetings during the strike, 32, 34, 35, 37, 39–40, 41, 44–45, 47, 48, 51, 149
Merchants, in the depression, 10; in the strike, 36–37, 48, 143
Methodology, 54
Middle class, associational memberships, 161, 166–168; clique memberships, 161, 168–170; economic relations, 187; merchants in the strike, 36–37; religious participation, 170–171; security, 185–186; social relations, 177–178; workers, 19, 33, 103, 165, 166–167, 169–171, 217–221, 222–224, 225–226, 228, 230, 231, 232–233
Ministers in the depression, 14
Mobility, through education, 183–185; through jobs, 79–80, 82, 86, 87–89, 160, 171, 175, 182–183
Morrison, Samuel Eliot, 54

National Boot and Shoe Manufacturers' Association, 47, 143; purposes, 124
National Boot and Shoe Workers' Union, 124, 125
Native workers. *See* Yankees

Newspaper, local, in the depression, 8, 10, 11, 12, 17; and factory managers, 23; in the strike, 32, 36, 38, 40, 44, 45, 51, 52, 138, 144, 147, 148; as upper-class opinion, 11, 17, 22
NIRA, and migration of industry from New England, 61–62; and National Boot and Shoe Manufacturers' Association, 124

Objectives of strike, as improved working conditions, 31–32; and increased wages, 31–32
Operatives. *See* Workers
Organized workers, 1–2. *See* Union; Strike
Owners, antagonism to, 19, 22; and cause of strike, 21; and control, 64; history of, 60, 64. *See* Managers
Owners, absentee, antagonism to, 19; and pay cuts, 22, 27; social distance from community and workers, 109, 114; in the strike, 5, 17. *See* Managers, absentee
Owners, former. *See* Managers, former

Pay rates, and age factors, 103; cuts in the depression, 13, 22; and ethnic workers, 98, 101; by department, 90, 94, 96, 98–99; as factor in the strike, 34; and management, 102, 142; and skill, 66, 80, 84, 90; and social stratification of workers, 103; and unions, 130; and women operatives, 85, 90–92, 95, 101, 103. *See* Wages
Picketing, 32, 40, 43, 44, 48
Piece work, and age factors, 103; and differential rates, 27; pay rates, 24, 26–27, 28, 85; and production rates, 104; and wages, 104
Planning, for industry, government, or society, 191; and human behavior, 191–192
Poles, and social class, 3; in the strike, 5. *See* Ethnic workers
Police, in the strike, 37–38, 44; as subordinating symbol, 43
Political organizations of workers, in vertical and horizontal extensions, 196; and social change, 188
Positional system, 161–166
Power, destruction by federal govern-

ment, 194; and economic and social change, 189; within federal government, 194; moving from small city, 187; of separate social systems, 195; in social hierarchies, 183, 188

President Hoover's Emergency Committee for Employment, 12

President Roosevelt, policy of spread work, 25–26; relief activities, 16. *See* Government, federal

Prestige, of craftsmen, 87, 97; and economic and social change, 189; of ethnic groups, 96–97; moving from small city, 187; of shoemaking under absentee owners, 119; in social hierarchies, 183, 188; of technological jobs, 67, 80, 83, 84, 85, 105

Profit-making logic, 107, 172–173, 175, 179; and wage cuts, 27, 30, 106

Protestant church, memberships of shoe workers and three lower classes, 170–171, 218–220; and relief activities, 16

Public opinion, as control of business, 187; in the strike, 31, 32, 52. *See* Newspaper, local

Public welfare in the depression, 14–15

Relief, coördinated, 14; in the depression, 13, 14; federal, 14–15; private, 14; public, 13–14

Religious participation of shoe workers and three lower classes, 161, 170–171

Retail sales, and absentee-controlled factories, 116–117; chain store, 27; and piece rates, 26–27; and profits, 25; and style changes, 26–27; and wage cuts, 25, 63

Risk-taking, reduced, 192, 193

Ritual, during strike, 32, 41; secular, 41

Riverbrookers, in the depression, 10; and ethnic prejudices, 93; and management, 142; and social solidarity, 93, 95; in the strike, 5; and union organization, 29. *See* Lower class

Sanctions of arbitration, 194

Savings, and the depression, 10

Schools, job training, 17; and mobility, 183–185; in status hierarchy, 188

Security, of middle class, 185; of upper class, 185–186; of workers, 18–19, 24, 61, 79, 83, 153, 157, 177, 180; of Yankee City, 178

Self-respect, and relief, 15; and technological jobs, 83

Shoe business, 4, 138

Shoe factory, and absentee owners, 17, 108, 121, 152–153; and delegation of authority, 71–72; external organization, 112–113, 115–118, 120, 197; in hierarchy of statuses, 67–68, 76, 110, 127; horizontal extensions of, 121; internal organization, 111–113, 115–118, 197; and leasing machines, 62; policy in the depression, 9; position in the community, 1, 35, 115, 121, 160, 197–199; processes, 61–62, 200–216; and social control, 67, 107, 109, 113; social integration, 67, 120; social structure, 60, 67–68, 82, 180, 197–198; spasmodic operation, 26; in the strike, 34; stock policy, 26–27; and style changes, 26–27; vertical extensions of, 121, 157; wage policy, 23. *See* Owners; Managers; Shoemaking processes

Shoe industry, in the depression, 63; history on Yankee City, 2, 56, 59–65, 74, 130, 135; migration from New England, 62–63; and National Boot and Shoe Manufacturers' Association, 124; and spread work, 9. *See* Horizontal extension; Vertical extension

Shoe Machinery Company, royalties, 25, 62

Shoe Workers' Protective Union, 28, 33, 34, 43, 44, 48, 125–126, 148, 149. *See* United Shoe and Leather Workers' Union

Shoemaking processes, assembling, 200, 208; cement and lockstitch processes, 61; cutting, 203–206; designing, 215; finishing, 213–214; hand and machine, 60–61; heeling, 25, 201, 211–213; and job leveling, 81–82; and job prestige, 84; lasting, 27, 60; making, 201, 208–211;

and mechanization, 76; packing, 201, 214–215; and rates of pay, 84; shipping, 201; and skill, 74–76, 81–82; sole-leather work, 200, 202–203; stitching, 26, 200, 206–207; and style changes, 60–61; trimming, 26; "Turn," welt, and McKay methods, 61, 98

Shopkeepers, and class attitudes, 22; and striker sympathy, 44

Skill, 73–89; and attitudes of workers, 66, 105–106; hierarchy, 66, 74–76, 87, 127, 175, 182, 189; and industrial unionism, 125, 127; and job prestige, 84, 105–106; and rates of pay, 66, 80, 84, 90, 105; in shoe production, 64, 66, 74–89; and status, 66, 76

Social behavior of workers, compared with three lower classes, 159–160, 161, 166–171; and economic status, 160

Social change, and relations of power and prestige, 189; and social class, 181, 187–188

Social class, 172; and economic change, 181, 188; in the strike, 35, 40, 43; structure in the community, 161–166

Social control, of absentee owners over community, 121, 157; of community over factory, 108, 109, 118, 119, 122, 152, 153; of ethnic workers, 97–98; in shoe factory, 67. *See* Control

Social distance, in absentee-controlled hierarchy, 118; between community and absentee owners, 114, 119, 179; between operatives and absentee owners, 23–24, 42, 108, 110, 119, 179, 189; between top and bottom of society, 196; between unions and management, 122

Social history of Yankee City, 54

Social integration, of shoe factory, 67; and separate systems of government, labor, and management, 194

Social personalities, 164

Social science, function of, 6

Social solidarity, and bargaining power, 93–95, 101, 123; of ethnic workers, 92; of horizontal associations, 123; of manufacturers, 44; in shoe departments, 93, 94; of women workers, 92; of workers, 36, 41, 44, 45, 78, 92, 103, 106–107, 132, 178

Social status, of clam flats, 3; and food, clothes, and neighborhoods, 19; of Hill Street, 3; and income, 19; of local factory owners, 118; and relief, 14–15; and technological jobs, 17; of workers, 33, 103

Social structure, of American society, 194; and unionism, 177; of shoe factory, 60, 175, 197–198; shoe workers' participation compared to three lower classes, 166–171; of total community, 2–3, 159, 161; of workers during strike, 34

Social systems, American, 194; and other, 194–195; and human feelings and sentiments, 192; and planning, 192–193; and vertical and horizontal extensions, 193

Social values, of occupations, 18

Spread work, and management, 25–26; as policy, 25; as workers' grievance, 24–25

Standard of living, and wages, 24. *See* Social status

State Board of Arbitration and Conciliation, 50, 145–146, 149; and arbitration of the strike, 52; recommendations, 51

State Labor Board, 31, 50

Status, interests of workers and management, 193; of old and new managers, 150–158; system, 3, 188; of workers in the community, 159

Strategy of strike, 32, 34, 35, 36, 40, 42. *See* Meetings; Ritual; Symbols

Strike, attack against management, 3; causes of, 1, 6–7, 21, 53, 54, 171–172; demands, 32, 47, 131; and the depression, 8–30; end, 52–53; and former managers, 134; history of, 1–2, 28–30, 31–53; and managers, 145; and mediation, 31, 47, 50; meetings, 32, 34, 35, 37, 39–40, 41, 42, 43, 44, 47, 48, 51, 52, 149; objectives, 31–32, 35, 36, 43; negotiations with managers, 134; picketing, 32, 40, 43, 44, 48; and public opinion, 31, 32, 143; as social crisis, 1; strategy, 31–32, 34, 35, 36, 40, 42; and workers'

fears and antagonisms, 14, 16–17; and worker solidarity, 36, 42, 45
Strike breakers, 29
Subordination of workers, by blocked mobility, 160; by division of labor, 160, 173, 175; and group solidarity, 160; by technology, 60
Supervisory controls, in shoe factory, 67–68, 69
Symbols, of management, 190; of solidarity, 32, 34, 41

Technological controls, in shoe manufacturing, 67–68, 70; as skill, 74
Technological development, 89; and expanding market, 60; and international economic organization, 195; and labor costs, 60–61; and larger economic structures, 190; and subordination of workers, 60; and union, 29; and worker insecurity, 180. *See* Mechanization; Shoe factory
Technological jobs, 68, 74, 109; and craft unionism, 176–177; evaluation by management, 67; and by workers, 66; and increased women employees, 91; and industrial unionism, 125, 127–128, 176; and personnel, 83; place in business structure, 67; and self-respect, 83; in shoe factory, 200–216; in skill hierarchy, 66, 74–75, 76, 80, 83, 175–176; and social prestige, 67, 79–80, 83
Townspeople, attitude to strike, 143; in the depression, 8. *See* Citizens
Tradesmen, and the strike, 6. *See* Shopkeepers

Unemployment, in the depression, 8, 12, 16, 25; as local problem, 12–13; in the strike, 29–30
Union, 108, 131, 132, 182; activities, 182, 184–185; as cohesive social group, 41; and collective bargaining, 70; and community, 6; control over workers, 31, 33, 37, 40–41, 42, 43, 132; craft, 65, 125, 127, 128, 176; as horizontal extension, 122, 123–124, 132, 179; industrial, 53, 65, 125, 126, 176; internal hierarchy, 177, 193; versus management, 7, 28–29, 31, 33, 122; objectives, 37, 40, 43; as social institution, 182; strategy during strike, 31–32, 33, 43–44; symbolic significance, 131; and wages, 5, 50, 130; and working conditions, 50. *See* National Boot and Shoe Workers' Union; Shoe Workers' Protective Union; United Shoe and Leather Workers' Union
Union leaders, attitudes of, 1–2, 6, 32; as control, 154; and former efforts to organize, 28; and management, 4, 144, 145; power, 193
United Shoe and Leather Workers' Union, 125; function, 130; organization, 128–129; purpose, 126–127; and technological hierarchy, 127
Unity of workers, 33, 34
Upper class, and antagonism of workers and lower class, 10–11, 17, 19–21, 24; associations during strike, 35; attitudes to lower class and to technological jobs, 18, and to strike, 4; in the depression, 9, 11, 12, 17, 18; economic relations, 187; and economic system, 22; "Old Families" and "New Families," 3; prerogatives, 18; security, 186. *See* Social status; Hill Streeters

Vertical extension, 132; of economic institutions, 195; through labor unions, 108; through manufacturers' associations, 108, 124; of shoe factory, 115, 121–122, 157, 178–179; of social order, 193; and the strike, 121

Wages, and buyers' market, 24–25, 63; and cost of living, 24–25; in the depression, 9, 13, 17, 23; piece rates, 26–27; and social status, 19; and spread work, 25–26; and the strike, 4–5, 24, 31–32, 35, 40, 43, 53; and supply and demand for labor, 98–100, 101–102; for women, 18, 24, 33, 44, 90–92, 101; and worker solidarity, 90–107. *See* Pay rates
Weatherby and Pierce shoe factory, 32, 140; and ABC Company, 140; and former strike, 29; and piece rates, 27; in the strike, 38, 52;

and workers, 41, 138. *See* Land, John

Welfare organizations in the depression, 13

Wheeler, G. C., 54

Women workers, attitude to work, 91–92; and differential compensation, 24, 90–91; in factory jobs, 18, 91; and mechanization and simplification, 91; and solidarities, 92; in the strike, 44; among workers, 33

Workers, and associational memberships, 161, 166–168, 218; attitudes to mechanization, 25, 60, 66, 80, and to spread work, 9, 25–27, and to the strike, 4; and to women's jobs, 91–92; and cause of strike, 21; as citizens, 160; cliques, 94–95, 103; clique memberships, 161, 168–170, 217; committees in the strike, 35, 42, 52; conflict with management, 1–2, 6, 28, 66–67, 109, 131, 172–173, 181; and department control; in the depression, 9; distrust of absentee owners, 9, 23–24, 27, 38, 42, 117, 177–178; ethnic, 225–226, 227–235; family structure, 161, 222; fears and antagonisms, 14, 18–20, 21, 22, 24, 25, 27, 142; grievances, 24, 25, 27–28, 182; as heterogeneous group, 33, 34; hierarchy, 182; history, 60; as industrial class, 160; and job evaluations, 66, 67, 83–84, 90, 175–176; and job mobility, 79–80, 82, 86, 160, 171, 182; and migration of factories from New England, 62–63; morale in the strike, 32; objectives in strike, 31–32, 63; and police, 43; political participation, 221; as powerful group, 32; prejudices to newer ethnics, 93, 97–98, 178; prestige, 79–80, 83, 176; and relation to machines, 76–78; relations with management, 9, 80, 81, 86, 88, 108, 114, 117, 130, 131, 172–173, 187, 190, 193; security, 79–80, 83, 158, 173, 176, 177, 180; and simplification of jobs, 174–175; social characteristics, 33, 161, 227–235; social life, 159, 160–161, 168, 177–178, 224; and social mobility, 171, 175, 177; solidarity, 36, 41, 43, 45, 78, 90–107, 132, 160, 171; status in the community, 159, 161, 171–172; strategy in the strike, 31–32, 34; subordination of, 160, 173, 175; and supervisory controls, 68–69, 80, 81, 173, 177; and technological developments, 60–61, 175–177; and unified demands, 8, 43, 47; and union, 31, 37, 40–41, 42, 43, 132, 176–177; -worker conflicts, 178. *See* Employer-employee relations; Women workers; Ethnic workers

Working conditions, improvement of, 31–32, 43; and union, 50

Yankee City Herald. See Newspaper, local

Yankee City, in American industrial life, 121; control over factories, 108, 109, 152, 155, 178–179; in the depression, 180; and industrial extensions, 132, 178–179; subordinate to absentee control, 121, 178–179

Yankee workers, 3; family, 224; memberships, 225–226; and pay differentials, 98; prestige, 96, 119. *See* Ethnic workers